BODY
THE ULTIMATE SYMBOL

Books by Olivia Vlahos

Body, the Ultimate Symbol
Far Eastern Beginnings
New World Beginnings: Indian Cultures in the Americas
African Beginnings
The Battle-Ax People: Beginnings of Western Culture
Human Beginnings

BODY THE ULTIMATE SYMBOL

OLIVIA
VLAHOS

J. B. Lippincott Company
New York

 For information address J. B. Lippincott Company, 521 Fifth Avenue, New York, N.Y. 10017.

FIRST EDITION
Designed by Ginger Legato

U.S. Library of Congress Cataloging in Publication Data

Vlahos, Olivia.
Body, the ultimate symbol.

Bibliography: p.
Includes index.
1. Body, Human (in religion, folklore, etc.)
2. Body, Human—Social aspects. 3. Beauty, Personal.
I. Title.
GT495.V57 391'.6 79-9137
ISBN-0-397-01368-X

79 80 81 82 83 10 9 8 7 6 5 4 3 2 1

To John

CONTENTS

PART THREE BODY RITE

ACKNOWLEDGMENTS

It is a pleasure to acknowledge the help of scholars and experts who have responded so graciously to queries or have discussed problems with me. Among them are David Beck, M.D.; Allison Brooks of Georgetown University; Roger A. Brooks, Capt., U.S. Army Reserve; Michael D. Coe of Yale University; Peter T. Furst of State University of New York at Albany; Irving Goldman of Sarah Lawrence College; and Gillette Griffen of the Princeton University Museum.

Thanks to Rena Gelissen for sharing a painful memory. Thanks to Fyle Edberg, Director of Library Services, Norwalk Community College, for making available the full resources of the Human Relations Area Files, and to librarians Thomas Gallagher and Dianne Flynn, who went far beyond the call of duty to help.

It was through the belief and encouragement of Beatrice Rosenfeld, my editor, that the book came into being. Bless her.

John Vlahos, poet and dramatist, has provided the unlimited benefit of a warm heart, a listening ear, and a critical eye. Bless him.

Our children, Michael Vlahos, Stephanie Vlahos, and Melissa and Marcus Ward, have generously shared their travel lore, their observations, and the perceptions of their age and time. Bless them.

PROLOGUE
Be All, End All

This is the story of the human body. Not of its anatomy, not of its becoming (for these have often been described), but of what its owners have thought about it, done to it, and projected through it.

In relating to the world outside, body is still our most reliable and trusty metaphor. We can hardly manage a classification or term (to say nothing of description) completely shorn of body references. The committee of the whole becomes "this august body" when we address it. There are the "world body," the "body of knowledge," the "body of the church," and the "corpus of a trust," which is merely Latin legalese for the selfsame thing. Admired individuals are the very "embodiment" of this virtue or that. The novice writer is instructed to "flesh out" his idea, to "give it body." We also put "body" into fabric, sauces, hair, and wine. And we certainly invoke body wisdom when we speak of "doing what comes naturally"—meaning without undue mental strain.

The body is what we came into the world with and what we leave behind. And in between there is only a growing awareness and understanding that starts with body and goes on from there. It was so in our human beginnings. Body dimensions gave us our first measure and our first grasp of the world. In body fluids we found our first color meanings. And through the tides of body life we learned cosmic rhythms.

On the body are expressed the secret wishes of the soul. We can adorn our bodies or forbid that vanity. We can pamper our bodies or punish them. We can deny bodily needs, disconnect with drugs, find through our bodies visions or oblivion. We may fancy ourselves as mind and as flame, pure and clear and clean. But mind is not heard, and flame is not seen. The messages of mind must be routed through the body, through its clumsy byways of muscle and nerve. And often, in one of its countless metaphors of feeling, value, hope, and dream, body *itself* is the message.

The specialist sees and appreciates even wider connections. Anthropologist Victor Turner has discovered body origins in much religious symbolism. Anthropologist Mary Douglas relates the many patterns of social life and the dynamic of social interaction to body metaphors. This excursion is through a smaller landscape, with ends less weighty and less profound. It offers a modest Baedeker, a guided tour through body meanings.

Vessel of life, the body is, as well, the ultimate vessel of meaning. And meaning, after all, is the beginning and the end of being human.

PART ONE

BODY ART

1

THE GRAPHIC BODY

Paint Me Human

Listening to the current ads for cosmetics, one is hard put to find the message. Lipstick is purveyed more often as food and drink ("Drenched with moisture!" "Try a lick of licorice!") than as agent of color. Foundation preparations are said to protect, smooth, nourish, and naturalize. Nobody mentions an improved, if artificial, complexion. And of color, not a word. Cheeks (and even eyelids, chin, and brow) are made to "blush" with substances that come close to brown if hue can be identified at all. One no longer hears of "rouge." That smacks altogether too much of color. And color is passé.

Forty years ago, the message *was* color, all color, and maybe more besides. Everybody had a familiar and affectionate anecdote about the painted face.

"Let me die with my boots on," said the Texas lady when told her end was near. "Hand me my lipstick."

"There must be some kind of makeup I can wear to bed!" wailed the new bride. It was a standard plaint when the naked face was the ultimate disclosure, to be postponed until comfortable habit had at last set in.

"You'll have to wait," announced the old lady to assorted young relatives. "I never leave the house without my corset laced tight and my rouge on just so." "Just so" nearly always meant two red stoplights positioned with geometric nicety but little regard for verisimilitude.

But that was then. Current wisdom would label such remarks as terminal vanity, the foolish conceits of an age now extinct as the dodo. And yet refugees from that other time read in the painted face something less superficial than fashion and more compelling than simple self-enhancement. Call it valor. Call it hope. Call it the personal endorsement of a collective ideal, the soul and its dearest wishes turned inside out.

Call it an art form, and one comes still closer to the truth. It was, after all, the first aesthetic expression of humankind and certainly the most enduring.

Our love affair with paint goes back a long, long way, antedating that other celebrated amour—Americans and automobiles—by four hundred thousand years, and probably more than that. It may have begun when, in the course of evolutionary events, retreating body hair first uncovered enough skin for canvas. "Charity begins at home"—so the old saying goes. And where else would art have had its beginnings? On walls? On bone? These were phenomena of more recent history. In their very first appearance, murals and carvings already show the expertise of long tradition—a tradition developed, it would seem certain, on the artist's own skin.

In those dim and distant times, what ideas could possibly have required graphic expression? This is what the reader will want to know. What messages were color-conveyed? One can judge only by the many known uses of body paint today and in the recoverable past. People have in this way represented beliefs about forces natural and supernatural, about the moral order, about spiritual and bodily health, about enhancement and enlargement both of the body social and the corporeal self. For starters, any or all of these would do. Take your pick. Lacking a time machine, we can never be sure of earliest purposes.

But we do know something about the colors themselves. They were, in the beginning, earth colors—the red and yellow ochers first of all. These contain iron oxides and are sometimes soft, sometimes hard enough to be chipped and shaped. Powdered, they mix well with fat, blood, plant juices, water. Heated, their red tones intensify. In addition to the ochers there was the black of manganese oxide or of charcoal soot, and there was the white of kaolin clay. In the long run, color itself may be the ultimate message.

The when of color

To find in an early hominid site the exotic earths—the ochers, clays, and oxides brought from afar—is to know instantly something important about the people who once called it home. They had interests that extended quite beyond today's dinner and tomorrow's sunrise. And those interests were expressed with color, if not in use then at least in appreciation. There is in such appreciation something that is akin to the power of speech or at least of the conceptual abilities that make speech possible.

The presence of imported ochers has been enthusiastically reported for the East African site of Olduvai Gorge (though not, it must be admitted, by the archaeologist of record, the eminent Dr. Mary Leakey). When and if such an appearance can be verified, the fascination with paint (and all that that fascination implies for developing mind) can be pushed back perhaps a million years or so, back to the time when humanlike creatures were transforming themselves into something better and something more.

We can be more certain of paint about four hundred thousand years ago when the Old World was peopled by *Homo erectus,* beings much like ourselves in looks and habits, though (judging by the size of their craniums) not quite so smart. In camps and habitations dating to those times, quantities of ochers have been found. One of the ocher sites is in and under what is now the city of Nice, France. It was apparently a summer home for many generations of hunters, who perhaps enjoyed the ambiance of the seashore as well as the fruits of the season. The kind of hut in which they sheltered from sea breezes, their hearths, their diet—all these can be reconstructed by the archaeologist. Preserved there is the footprint of a man who slipped in the mud. So is the imprint of a (possibly) wooden bowl. So are many ocher crayons, well rounded and quite worn away in use. Use on what? What else? On the skin.

Of time and use

A hundred thousand years ago reigned the Neanderthals, variants of our own species. Their uses of paint and even something of the meanings they attached to it can be more easily read. Lumps and "crayons" of color, the ochers and black manganese oxides, have been found in several locations, some of them rubbed as if applied to skin, others scratched as if for powder. Archaeologist François Bordes has uncovered in the cave of Pech de l'Azé, southern France, a limestone palette on which paints were ground. No murals have yet been found. These are credited to artists of the last Ice Age. Neanderthal paints must have figured mainly in a body art.

There are, however, a few painted artifacts. At Tata, a Neanderthal site in Hungary, there was discovered some ten years ago a plaque carved from a mammoth's molar tooth. Shaped to a rounded rectangle, edges painstakingly beveled, it is a beautiful thing to see. Now lustrous white, it was at one time rubbed all over with red ocher. This happened some fifty thousand years ago. Successors to the Neanderthals, people of our own sort, would so anoint little female figurines, which were the precious talismans of their time.

Evidence of ocher use by the Neanderthals comes from widely separated sites in Europe, Russia, Africa, and the Middle East. In some places they appear to have mined the pigment, adding it to shrines built to honor bear and deer. Neanderthals often buried their dead on lumps of red ocher or sprinkled it as powder over bodies tightly flexed. So did their successors, who added, as well, lavish gifts and goods in an escalation of mortuary display.

So did Ice Age migrants from Asia to the New World. An eleven-thousand-year-old burial in Montana, uncovered accidentally by a bulldozer, revealed the bones of young hunters covered, as were grave gifts of stone projectile points, with powdered red ocher. All across the continent we find such burials, made by people of somewhat more recent times. The skeletal remains of one such group in Maine have been dubbed the "Red Paint People," so lavish had been their use of ocher.

Red, black, white. Always these three. In burials. On artifacts. Across the walls of caves and cliffs when people later came to produce representational art. And on the skin. That last we must infer from the evidence in the earth and in the ways of the living.

The why of color

Like the ancients, living tribal folk have preferred red, black, and white. They have used these colors (often only these colors) in masks, in costumes, on wood, rock, hide, and on their own skins. Specialists have usually explained preference by way of limitation. If coloring agents are few, they reason, one makes do with what one has. And practices begun in scarcity are then enshrined in habit.

Anthropologist Victor Turner, whose interests lie in religious symbolism, thinks he knows another reason. If he is right, the colors of the ancients can themselves tell us something about *why* men painted, as well as *when*.

Ideas about the meanings of color came to Dr. Turner while he lived among the Ndembu, an African tribe of northwest Zambia. Here many colors are used, most of them named descriptively—"like this plant" or "like this flower." Only red, black, white have primary terms. Only these colors are used in celebrating the great events of life. For only these are power colors.

The Ndembu are quite precise about their meanings. White is milk and semen and therefore dual in sexual nature. White is goodness, purity, strength. It has the power of hunting and chiefliness and the ancestors. It has the power to cleanse, to make fertile, to make visible that which is hidden. Red, too, has power, the power of life and blood, the power of the bloodline, which is family. Black represents night, concealment, and death, both ritual and actual. Associated with elimination and decay, it encompasses suffering, disease, the evil of witchcraft and of sorcery.

Black as color, as meaning and feeling, appears wholly unconnected with the darkness of Ndembu skins. Which is to say, skin tones and the abstract meaning of black have nothing to do, one with the other. Yet women with very black skins are thought to be sexually desirable, passionate, fertile. This lets us know that, for all its connotations of dread, black has its own powers of life and love. What the body expels, after all, nourishes new life, and what dies is reborn. One remembers, of a sudden, the tomb treasures of Tutankhamun and the figures of the young king painted black with resin. Black for rebirth. Black for fertility. Turner reminds us that black is the color of rain clouds and therefore signifies life, especially in arid lands.

Red, too, is ambivalent in meaning. It can bring evil as well as good. If it is the blood of game and food, it is also the blood of birth and menses, to be feared and shunned. If it is the blood of life, it is also the blood of human killing, of murder, war, and power unrestrained.

The common message

As he traced in other social settings the meanings of the "primary" color triad, Victor Turner found variations on Ndembu themes but no significant differences. Wherever he looked there was enough similarity to suggest something close to universal accord. Even in our sophisticated modern world the color symbols appear, trailing their ancient clouds of meaning. We exchange red eggs at Easter; dress brides, novices, and candidates for confirmation in white; mourn in black. The pre-Mao Chinese preferred to emphasize in marriage not purity but new life, and so brides were dressed in red. Red clothed the bodies of the deceased, while survivors mourned in white.

This is not to say that other colors lack significance. One

thinks of blue: for us "true blue"; for ancient Britons a tribal color; for the Maya, priestly in nature. One thinks of yellow: Aztecs so painted the intended victim of sacrifice; yellow in Southeast Asia is the color of the monk's robe. There is purple: in Mediterranean lands the emblem of royalty; in some villages of southern Mexico, the color of the bridal costume. But blue, yellow, and purple do not everywhere carry the same or even similar meanings.

Red, black, and white do because, believes Dr. Turner, they are body colors, colors of the body products and fluids associated with the process of living. These were colors not merely perceived but *felt*. For they attended life's beginning, life's continuity, life's end and dissolution. In earth's body, in her ochers, oxides, and clays, beginning humans found metaphors for the great events of their own bodies and souls. Those metaphors remain, however elaborated and transfigured, because there remains for humankind a common body reality.

The message in action

Half a world away from Africa there live to this day the Tchikrin, whose ways have been studied by anthropologist Terence Turner. A jungle people, isolated, horticultural like so many other inhabitants of the Amazon Basin, the Tchikrin settle in villages. Each is constructed as a wheel of dwellings orbiting the central men's house. Here young bachelors shelter as they await marriage and permanent residence in their wives' maternal homes. To the central house come the married men, as to a club, for relaxation, companionship, ritual. And of ritual there is much. It marks all the stages of Tchikrin life, and marks them on the skin. Birth, initiation, marriage, death—all are announced in paint. Women decorate themselves, other women, their children, their menfolk. They may fingerpaint or trace delicate lineal patterns with wands made from the central ribs of plant leaves. Men paint only themselves or one another, and always with hands and fingers.

White is made from kaolin clay, which is the food of ghosts, the home of ghosts, and their ultimate representation. It is, says Terence Turner, a terminal color, with emphasis on its inherent purity and not on the generative properties with which it is imbued by other folk. White does not figure in the active transitions of life and, perhaps for that reason, is seldom used by the Tchikrin.

Black, made from the fruit of the genipa plant, is called *tuk*, the word for "dead." It signifies the cemetery area between village and forest. And just as that cemetery is a zone of transition between the human realm and that of animals, between the partly dead and the wholly forgotten, so black is a color of transition. Boys at age eight, taken from their weeping parents to enter the men's house, are painted black all over. Dead as children, they are to be reborn as men. Girls ready for motherhood are painted black on thigh and breast. For special occasions and for every day, it is the trunk of the body, the upper limbs, the cheeks and sides of the face that are blackened. Here, for the Tchikrin, is the home of the inner self, the seat of what we would call biological needs and drives. Painting these areas black, Turner came to understand, imposes the strictures of society over the ruthless and demanding self.

Black face is immediately covered with the waxy red made by boiling seeds of the urucu plant. There is no vanity in this act. It is a statement of human intention: to breathe life into the boundaries and render them tolerable and even creative. Always the body extremities, lower arms and legs, top of head, around eyes and upper face—everything swift of thought or movement—are painted red.

Seeing the naked painted Tchikrin, man or woman, one is reminded just how much canvas has been lost in acquiring dress, that final adjustment to life without hair and in cooler climes. We must make our metaphorical statements with clothing, with our costumes, uniforms, and various states of exposure. Yet, given a period of sanctioned nudity, we seem to long for protective coloration. The greater the allowable expanse of skin, the greater the haste of pale Americans to activate the melanin cells. If we lack time for the sun, we apply tanning lotion.

There looks to be nothing at all of nudity in the painted Tchikrin body. Indeed, it seems from a distance to be clad in tights and leotard, nicely accessorized with gloves and socks. The significance of reddened extremities may once have been widely shared in the Americas. We see sock-painted figures on pottery fragments found in ancient temple centers of the Mississippi Valley and coastal Peru. The shards depict men dancing, fighting, and otherwise cavorting, naked as jaybirds save for their painted socks.

Basic black

Painted socks and painted gloves are also seen among the Cubeo, a people of the Amazon Basin far in its northwest region. But the color of use is black, fertile black, black that conceals and protects. Above all things, the Cubeo value growth of the body: growth in size, muscle, fat; growth in sexual potency and power. In any situation that threatens fertility and growth, black is applied. The man chosen to "blow the river," to spray it with water and tobacco smoke, first takes care to blacken hands and feet. Protected, he may then enter the water, there to warn the anacondas and river spirits of people about to bathe, or to notify them of a baby's birth. Women apply genipa whenever they must handle tobacco and the fatigue-allaying coca leaves, ritually dangerous and reserved for men. So we are told by anthropologist Irving Goldman, who has studied the Cubeo.

Sometime during his first ten days of life, the new baby is painted black to "remove the fetal skin," which will hamper his further growth. And when he is a year old, he is painted black again to receive from his grandfather a name. Before the old man are placed two calabashes, one of genipa dye and one of milk drawn from the mother's breast. Dipping a finger in each container, Grandfather touches the baby's heart and his nose just where it joins the forehead, at just the spot from which, at death, the soul will emerge and escape.

The ambivalent meanings of black are acknowledged by many other folk. When head-hunting Jivaro of the Ecuadorian Montaña were still following those customs for which they are best known in the modern world, they wore black in war in order to display the ferocity of demons. They also wore it when they went courting, to lend extra power to seduction.

North American Indians of the woodlands also had mixed

feelings about charcoal black. Worn on face and body, it announced a state of fasting—the "little death" of self-denial which awakened the pity of animals and ensured the hunter's success in the food quest. Back in the wigwam, his children seconded his efforts by a similar blackening. Black likewise awakened the pity of spirits who could reward one's hungry vigils with the great visions that promised glory in life. And there was in black something more than an announcement of condition. It had the active power to sharpen second sight. The eye well rimmed with charcoal could bring into focus even the most vacillating spirit.

Rwala Bedouins of the Middle East had much the same view of black (and maybe still do). But it was not so much second sight that counted as the real thing. In other years they have been observed carefully blackening the rims of their binoculars. This may represent a transfer from the eye itself and from the sooty kohl with which the eye was habitually encircled. One begins now to think the custom was originally intended not so much for cosmetic effect as for sensory utility. Perhaps this use of black as vision enhancer may help explain the fondness for eyeliner that has characterized Middle Easterners of both sexes for nigh on five thousand years.

It is fashionable still in India. Feminine makeup emphasizes the eyes, never the mouth. The young girl about to be a woman dons eye makeup the way our little girls twenty years or so ago affected lipstick. Baby brother may be blackened too (certainly if the family lives in rural isolation), but for different reasons. He is smudged all over and dressed in rags to confuse and foil envious spirits who love to rob a family of its treasure.

Terminal, cardinal, or clean

Depending on geographical location, native North Americans now and then chose red or white as colors of the vision quest—with somewhat different connotations in view. White, by and large, appears to have carried a more explicit charge of the sacred. Creek priests were painted white. So, frequently, were candidates for self-torture in the Sun Dance, that most characteristic and most sacred ritual of the Great Plains.

Among some African groups, white is used to coat the bodies of the dead, signifying purity and the ascent to ancestorhood.

In Zaire the faces of frightened Suku boys awaiting initiation are painted white, as are the sacred wooden masks worn on this occasion. Perhaps it is the ancestors who are here invoked. Perhaps through paint an identification of boy and spirit is achieved.

One can only wonder whether it was by original association with the spirits that whiteface achieved its worldwide popularity. It may be that the connection was of a more secular order. White skins have everywhere announced more time indoors, a certain consequence of rank and wealth. Doubtless, paint gave the upwardly mobile a useful counterfeit. Whatever the original stimulus, chalk-white faces on skins of decidedly darker hue have represented essential elegance to ancient Egyptians, Sumerians, Cretans, and sundry other folk of historical reputation right up to and through eighteenth-century Europe, where, it was said, the white lead which provided instant pallor produced a rash of skin cancer. In the Far East, rice powder whitened the faces of Manchu ladies of high degree, Japanese geishas, and Kabuki actors. One can still see the fashionably white mask among Philippine boat people, painted on the faces of young couples ready to tie the knot.

White for Australian Aborigines connotes water and the purity of water. It is also used in war, carrying thus a strong whiff of holy intent. Certainly righteous anger is demonstrated among the Tiwi of Melville and Bathurst islands just off the north coast of Australia. Here, as among most Aboriginal groups, might does not make right; age does. Older men of renown have many wives. Young men have none and are therefore frequent poachers in their elders' preserves. Apprehended, the young culprit gets his comeuppance in a duel which is more judicial drama than physical contest. For the indignant elder is the inevitable winner, made so by the support of his peers and even by the relatives of the accused. To the duel, the plaintiff comes painted all in white, wrathfully chewing his white-painted beard or its substitute, a ball made from goose feathers.

The fire of life

For ritual, Australian Aborigines prefer red. Sacred artifacts are kept well rubbed with ocher (as anciently the Neanderthals anointed their ivory plaque). On their bodies men paint designs in blood or ocher, which is the blood of the Dream Time, the blood of ancestral heroes who lived then and live again when the Dream Time is painted into being. In paint is drawn the dichotomy between sacred and profane. It is a boundary which in our world appears by way of clothing, in the vestments of specially trained religious practitioners. But in simpler social settings, knowledge of religious symbols belongs to all adult men (seldom women), and religious art, body art, is intended less as costume than as life.

Red for the newborn comes close to being a universal prescription. This is the color with which a Cubeo infant, immediately after birth, is separated formally from the animal world and made human. His mother girds him with decorations no animal can make: a bracelet of twine, a necklace of seeds. She paints his body red, and on his face, ever so carefully, designs a rash of red rosettes. In this way he is made to look rather like a jaguar baby. Yet similarity in this instance commands distance, commands protection from the jaguar, terrible in itself, more terrible still as the alter-ego of the sorcerer.

The red and the black

Some tribal groups redden the new parents along with Baby, and the Siriono are one of them. Ceaselessly they travel the Bolivian jungles in search of game, ever more scarce as the years go by. Their intense hunger for meat is expressed actively, vocally, and in the language of color. Red is the color of females (as well as new babies), through whom are traced the Siriono lines of descent. Fire, too, belongs to them, given long ago by that quintessentially "red-hot" animal, the jaguar. To females belong the animals forbidden as food, and theirs are the chores of the hearth.

Red is opposed to black, which is the color for men, for hunting, for water, and for the edible animals associated with water. Men hunt animals and with them are enabled also to "hunt" women. Sexual favors are traded for meat, and never mind who brings home the bacon. Hunting ability and sexual potency are one and the same, and the successful provider is sure to have an active list of wives and mistresses. His meat, nevertheless, goes into female fire, where it is cooked. Perhaps the giver of the meat is "cooked" as well, suggests anthropologist John Ingham; domesticated, bound.

Among the Sharanahua of Peru, exchange has been elaborated into ceremony involving costume, song, challenge and response, mock flight and mock capture, rewards both public and private. The desire for meat appears, in this setting, less a matter of real hunger—Sharanahua women do cultivate manioc and plantain—than the thrill of the chase. So thinks anthropologist Janet Siskind, who has lived among them.

No gold digger wheedling diamonds was ever more expert than the Sharanahua woman wheedling meat. And no big spender temporarily out of funds was ever more thoroughly chilled than the Sharanahua hunter out of game. "Let's eat penises," a scornful woman will remark to her equally scornful cowife, making sure their comments carry to the luckless male huddled disconsolately in his hammock.

When Sharanahua women as a group put on their finery and begin to sing, the men groan or exult, depending on who will be sent out to hunt for whom. Adultery, after all, is the name of the game, and jealousy is very bad form. For the "special hunt," women paint their faces and legs with utmost care, drawing on an undercoat of red achiote with delicate patterns of black lines. Men may paint themselves occasionally but never with much interest or skill. For Sharanahua women, paint enhances the scenario of seduction and the battle of the sexes. Men are expectably snared. "She was beautiful," a man will say of a woman. "Her paint was lovely."

More color scenarios

> I believe in this part of South America, man exists in a lower state of improvement than in any other part of the world . . . their hideous faces daubed with white paint, their skins filthy and greasy, their hair entangled, their voices discordant, and their gestures violent . . . one can hardly make oneself believe they are fellow creatures and inhabitants of the same world.

So Charles Darwin described the Yahgan of frigid Tierra del Fuego as he saw them in 1832 while voyaging on H.M.S. *Beagle.* Vision, it must be said, was edited by his own culture. He was deeply offended by Yahgan nakedness but, more than that, by their nakedness in the intense cold. Observing the grease, he failed to take into account its insulating properties. The paint he found merely savage. How would he have reacted, one wonders, to the Yahgan color quest?

Regularly they went on pilgrimage for the precious red oxides, which were afterward cooked, dried, and stored carefully in a sea lion's gut. White chalk, nearer home, had to be mined. Thoroughly chewed and mixed with saliva, it made a satisfactory paint. Powdered charcoal mixed with fish oil made the necessary black.

Everybody painted. Only old men were negligent in this respect—so says Father Martin Gusinde, the Jesuit priest who studied the Yahgan in the early 1920s. Each of the moods of life had its proper colors: red for happiness, black for grief and mourning, white for ritual. And the designs in these colors announced just what event, blessed or otherwise, had elicited the mood. Even the cause of death was portrayed on the face of the corpse. Cheek and jaw of the drowned person, for example, were painted with a black-and-white mix, the gray of sea-floor sediment, while goose-feather down over the forehead signified the foam of waves. As it happened, men were frequent victims of drowning, for although Yahgan women swam, men did not. Such was the division of labor and of skill.

When one visited what Darwin called the "haycock" home of a Yahgan friend, etiquette required white paint with a red overlay. The host painted also if he had received warning from the approaching canoe and so gained time for the amenities. Otherwise, he excused himself sometime during the visit and repaired the omission. To remain unpainted would have been unpardonably rude. Here on survival's very edge, in the "lower state of improvement," we see the equivalent of dressing for dinner, however rustic the surroundings and straitened the circumstance.

For the Yahgan, paint affirmed the social enterprise. No one ever joined a group, particularly when there was singing, without applying color. Lumps of pigment passed from hand to hand and from hand to face as, in a gathering of our young, a "joint" might be puffed and shared.

It is not so much the social as the common religious enterprise which was affirmed in paint by people of the Andaman Islands, a chain stretching from Sumatra toward Burma. For them, red was warmth, life, energy, force. White was the color of the spirits and the joyousness of bright, fine weather, of delicious honey, which was the gift of fine weather. After each meal, red paint was applied in lines and white clay in zigzags, recalling the markings of certain snakes. Thus they thanked the animals of land and sea, their only source of sustenance, by painting imitative designs on grateful faces.

In the early 1900s, when the British anthropologist A. R. Radcliffe-Brown lived among them, the people of these islands were declining in numbers. Certainly in decline were those who had contact with the outside world and its unfamiliar diseases. But there were then and there still are Andamanese who defend their privacy with arrows, who have not been studied, and who cannot be seen. They may still paint for religious reasons, after the fashion of older times.

Girding for battle

When a woman of our world says she is "putting on her war paint," she seldom has combat in mind, and certainly not a blackened face. In very many cultural settings, men mean just that. The ancient Maya associated black with war, death, and the underworld. Yąnomamö of the Orinoco Basin are still raiding one another with enthusiasm and still painting black for the occasion. Charcoal is thoroughly chewed to make a blue-black pigment which, applied to the face, assures the wearer of an appearance of ferocity he may not altogether feel.

Also painted black for battle are the Dugum Dani, a people of the Western New Guinea highlands now called Irian. They prefer, however, to mix charcoal with pig grease, which moisturizes while it decorates. Mourning requires the weakening action of mud, which dries the skin. Black-laced fat is also used for what Karl Heider calls profane art, figures drawn on rock walls. Red ocher, used on the body only for the initiation of young boys, is prescribed for sacred drawings made on cliffs hard to find and difficult of access.

Other people of New Guinea run riot with colors, some of which are now pigments produced by modern chemistry. There is maintained nonetheless a sexual division of hue. Though red is a magical color and white an emblem of fertility, nothing is quite so impressive for great occasions as big, bad black, the color of war and of dominance, the color which submerges the self in an ancestral soul. It is therefore a male color, the essential background on which bright highlights may be emblazoned. For ceremony men vie in creating gorgeous fantasies of paint, foliage, shells, and feathers of the rare and exotic birds for which New Guinea is famous and which human costume often mimics. Human males, like male birds there, monopolize display. This is not to say that women are forbidden to paint themselves or one another, predominantly in red or white. They do, but often more as a means of advertising the renown of father or husband than as personal enhancement. Even then they are never more than faint echoes of male glory. Like the birds.

The painted American

It is the female bird of the civilized world who has been accustomed to bright plumage, and facial paint has been an art form sex-linked and nearly exclusive. Since the beginning of written history and the appearance of the state, this has been so. From time to time men have affected cosmetics. The fierce warrior of seventh-century Japan painted as cheerfully as his graceful wife. Eighteenth-century gentlemen of the French court powdered their faces as they powdered their wigs, and applied beauty marks with aplomb. But these have been sometime phenomena, and in recent years artifice in both sexes has been in retreat.

We appear, however, to be on the verge of a renaissance in the decorative arts, especially those that require the use of face as canvas. Blue-jeaned damsels who until very recently would never have compromised with paint their sincerity or liberation are now beginning to add dabs of this and dabs of that. Eyes mascaraed, everything else pristine. Rouge on chin, nothing more. Or perhaps a bit of lipstick with more shine but barely more color than petroleum jelly. These tentative additions appear to convey a delicious sense of daring, even of sin. The devotee of the Natural can now be observed to linger more moments than are strictly necessary, preening before her mirror.

She is not always alone. If in the renunciation of paint the sexes became more truly one, oneness now persists in its readoption. Makeup kits for men are becoming marketable items, along with a bewildering assortment of baubles, bangles, and beads intended for masculine adornment.

If one looks hard, one can even discover among men a symbolic as well as a decorative use of paint. And where should one look for that but in the very cockpit of contemporary ritual, the sports arena. Consider the boxer, laced by his solemn trainer into gloves and, at the last moment, daubed cheek and breast with grease. "It deflects the blows," explains his second. Perhaps. One remembers the Dani warrior greasing up before the battle plunge.

Consider the football player in uniform exaggeratedly male, pants tight as a coat of paint, tight at the knee to bunch thighs and calves, tight at the crotch to outline sex. And this juggernaut of masculinity adds the final touch: a smudge of black paint under each eye. "It reduces the glare," says the coach. No, sir. It is paint for battle, signifying both resolve and unity for men who will labor in this, the "little brother of war." So the Creeks labeled their ball games, and there is little reason to alter that perception.

In war itself and among the men trained to its tasks, the use of paint is still more to the point. Certain elite battalions—rangers, commandos, and other special forces organized during World War II—make much of black pigments. "Camouflage," say the commanders. But notice that the rangers charcoal their faces—ever so carefully, allowing not the smallest fleck of tan to show—often when neither attack nor infiltration is anticipated and always for the camera's eye. The blackened face is a better earnest of intent than the beret.

Yes, paint is back in business and back on the skin, doing what it has done from the beginning, since first humans found in oxides and clay their body metaphors. Paint binds and separates, defining the pairs of opposites: sacred-profane, male-female, sick-well, good-bad, high-low, living-dead.

Paint sets human beings apart from nature, while at the same time putting them in touch with natural forces apprehended but dimly understood. The closer to their environment a people live, the more dependent they are on its bounty and the more subject to its terrors, why then the more important, it seems, is this dual affirmation. "I think, therefore I am," said Descartes, the French philosopher. The Yahgan fisherman might have put it otherwise: "I paint, therefore I am."

Paint announces the crisis times of life and announces them on the bodies of individuals in stress. By the power and meaning of color, feelings are channeled and vulnerability protected. By color are companions notified to give support and consolation.

In paint a community displays collective values and sustains social commitment. By looking much like everyone else, the individual demonstrates participation and belief in a common identity.

Finally, paint communicates on the outer skin something of the inner self. It is a self enlarged, idealized, enhanced. And the soul rejoices and is content.

Paint me as I long to be and need to be and am. This is the message of the graphic body. *Paint me brave. Paint me charming. Paint me clever, able, wise. Paint me eager to please but to mine own self true.*

Paint me human.

2

THE PLASTIC BODY

Improving on Nature

In the Book of Genesis it is written:

> So God created man in his own image,
> in the image of God created he him;
> male and female created he them.

Not altogether in his image, it would seem. For Adam was created uncircumcised and therefore "blemished" in the sight of the Lord. Not until much later was this information communicated to Abraham, who set about to remedy the defect, and all his progeny after him.

The Hebrews were not alone in the quest for physical perfection, whatever that was accounted to be. Never has there been a people on earth willing to make do with what comes naturally.

Like a rough-hewn stone, the body has been perceived as unfinished, in want of the carving, perforating, incising, refining, and realignment that would make of it a thing of beauty and a joy forever. Chassis imperfections have never been considered fully remediable with a mere coat of paint. Something more drastic and more permanent has always been decreed, something requiring the sculptor's eye and the sculptor's skill. What Nature had been able only to begin, the human hand would bring to perfection, would complete.

When one considers the pain involved in body retooling, one is hard put to understand the human eagerness with which it has been sought. Men, women, and children in all places and through all stages of history have rushed to have themselves completed. But then, what is pain to the glory of the realized ideal? What is pain to the supreme satisfaction of keeping up with the Joneses? For, like the uses of paint, bodily readjustments which begin as philosophical statements invariably end as fashion.

Beginnings of plastic art

Just how long ago humankind felt the urge to improve on Nature is anybody's guess. Skeletal remains reveal nothing of redesigned soft parts, edited features, and various other tinkerings. Doubtless the expertise accumulated through millennia of chipping stone tools, butchering prey, and processing hide readied the prospective sculptor for his task. It is probably also safe to locate the art of body remodeling in a time following the discovery of body painting. Indeed, it may well have been something of an adjunct. Just as fashionable women a generation ago joyfully abandoned curling irons in favor of the permanent wave, so, too, our ancestors may have adopted the permanent color that tattooing can provide. Merely pierce the skin, lift it, rub coloring material well into the wound, and there it stays, an indelible announcement of status, membership, or beauty. This is not to say that body painting was then abandoned. If our forebears were at all similar to their living descendants, they continued to apply paints as messages of enhanced feelings or guides to social modes. But such messages are changing and transitory. The tattoo is forever, a cultural stamp of approval on the finished product.

Some of the rare carvings and engravings made of the human figure during the later years of the Old Stone Age, perhaps as long as twenty-five thousand years ago, show geometric markings on face and form, markings which may represent tattoos. They may represent painted marks as well; there is no way of knowing. By the Neolithic, when the world's people were fast adopting farming ways and making the clay bowls and jars that can be accumulated when one can count on a permanent cupboard, tattooing was definitely in vogue. Ceramic figurines everywhere offer proof. One of the facial designs favored in the Middle East—an inverted trident in blue—can be seen to this day adorning the chins of Berber women of the North African Atlas Mountains. The design is said to be a vestige of an ancient, forgotten script of the region. One suspects an even earlier legacy. The pretty blue marks are skillfully enhanced by the Berber girl's fondness for kohl-rimmed eyes with large black beauty dots at each corner. Round red cheeks complete the picture. As to which decorative convention came first—tattoo or paint—the coquettish woman has not a notion.

Permanent paint

The sequence is, however, well understood in New Zealand among the native Maori. One of their myths describes how the great hero MaTaora, while braving the underword like Odysseus, learned tattooing from his dead father-in-law as a way of making painted designs a permanent fixture. Thus Maori tribal pride gave to common knowledge a local origin. Tattooing with an adzelike bone instrument was long in use throughout Polynesia and for both sexes. It must be said that, while men were more extensively and visibly tattooed, women tended to choose the more exotic, out-of-the-way locations for designs—the tongue and genital region, for example. On some islands the fully tattooed body was in order. Others favored an analogue of walking shorts. On still others, tattooed tights were preferred.

The Maori concentrated their unique designs on the face, preferring rather to carve them into living flesh than to pierce and tint. It was an ordeal so painful that a man usually had one side of the face carved and rubbed with soot in his early manhood and planned to complete the operation later in life. So swollen were the mouths of sufferers that they could receive nourishment only by use of special funnels provided for the occasion. Because of the bloodshed involved, the operation itself was one fraught with spiritual as well as physical danger. Even the plot of land on which a man of high rank had been carved remained taboo for generations after the event.

Beautifully carved heads, on or off their owners, were highly prized. The severed variety was preserved by steaming and smoke-drying. So successful was the technique that in the early 1820s, when Europeans were in the area, the Maori did a brisk trade in tattooed heads. In order to satisfy demand, they eventually took to carving the faces of their slaves and then removing the heads afterward. Some slaves, it is reported by Elsdon Best, were so little conscious of the honor done them that they managed to decamp with head and body in one piece.

Among many other peoples the tattoo is an insignia of rank. By tattooing the family crest on every available and visible patch of skin, those Northwest Coast Indians having social pretensions made sure that they could never be mistaken for lesser folk. For the warrior with minimal uniform, the tattoo can serve as a permanent equivalent of service ribbons and battle stars. The Kalinga of the Philippines, with no guns to notch, added the appropriate tattoo for each head taken as a trophy.

Whether the privilege of tattoo is assigned to men or to women often seems of less importance than the sexual division ratified thereby. And even when both sexes are so decorated, the designs preserve the elemental distinctions. No unisex here. And no whimpering, either. Among girls of the Indonesian island of Alor it is excessively bad form to speak of pain during the operation or at any time thereafter. One pays the operator and offers polite thanks. After all, think how gorgeous one is being made.

For Ainu women of the northernmost Japanese island, the tattoo was like a wedding band, applied after marriage to hands, brow, and lips. The dutiful woman, so the markings seem to proclaim, worked only for her husband, thought only of him, spoke only for him. The last injunction required lip markings absolutely unique in the annals of tattoo. A sort of larger mouth, curled in a permanent smile, was made to surround the woman's natural lips. Nobody was quite sure in the early 1900s just how or under what circumstances the custom had arisen. Certainly it was a very old custom, called as it was "blackstone wounds," a term recalling the use of obsidian knives. European observers, noting the luxuriant growths of facial hair sported by Ainu men—Caucasoid rather than Mongoloid in background and therefore very hirsute—concluded that the lip tattoo provided for women a surrogate mustache.

It may be that the Japanese, who came late to these islands, picked up tattooing from the aborigines. Despising the Ainu, they made the tattoo a brand of shame with which they marked the traitor. By the sixteenth century, however, meanings had changed. Japanese prostitutes were tattooing the names of favored customers on their thighs. And when sumptuary laws went into effect, when costly clothing was forbidden, the ardent social climber could hide under his somber tradesman's gear a tattooed body as elegant and costly as a brocaded gown. To this day in Japan, the tattoo is an occupational emblem for firemen, carpenters, and gangsters. And it remains an exquisite art form.

The engraved body

For one of very dark or black skin, the permanent paint of tattoo is a useless conceit. But the associated puncturing and incising will produce keloid welts of quite spectacular appearance. The aesthetic effect of wounds rubbed with common dirt was discovered in Africa at least as long ago as eight thousand years. Figures thought to have been painted then on Saharan cliff walls bear scar patterns not too different from those to be seen today on living West African skins.

Africans, and to a lesser extent peoples of Australia, New Guinea, and Melanesia, scar enthusiastically and for a number of reasons. There are tribal scars ("Only True People mark their faces in this manner; I am a True Person"). There are clan scars ("We are members of the Peanut Clan, children of a human woman and the Spirit of the Peanut; can't you see how wrinkled we are?"). There are maturity scars, gained through initiation ordeals ("The Bush Monster chewed me up and spat me out again.

See here the marks of his teeth!"). There are the scars of personal enhancement ("What a charming girl! What a pretty pattern of scars. Like a constellation it is! How well it suits her!"). And the scarified little minx, so flattered, preens and arches her back, resolving to wheedle Dad into financing yet another line of scars just here around the navel. Fine feathers make fine birds. So does scarred skin.

Retooling the nether regions, male

Genital scarring for both sexes has little to do with vanity, since the effects are usually hidden from view. But it has a great deal to do with the inner self. So ancient is the requirement of circumcision (everywhere lingers the memory of the stone knife but recently discarded), so widespread is it in the world (only Eurasian peoples are exempt, and not all of them), that scholars have sought long and earnestly for beginnings. Could the operation somehow promote fertility? Probably not. Could it have served the interests of hygiene? Considering the septic settings of the rites, probably not. Could it be a substitute sacrifice—a part of the body instead of the whole? If so, the world would long ago have lost its generational complement of young men before ever the substitution could have been effected.

No, held the *Jewish Encyclopedia* of 1901 (whose authors understandably subjected the phenomenon to its most penetrating analysis), the act is everywhere associated with initiation, with the continuity of life, with the completion of man by the hand of man. That the Hebrews saw it as a covenant with God and circumcised the male baby on his eighth day of life was a unique interpretation of what seems to have been originally a celebration of puberty.

The earliest anatomical evidence of the practice is to be found among the Egyptians. Results can be seen on mummies of the second millennium B.C. Historical evidence is older still. The Narmer Palette, a testament of conquest as well as a grinding stone for cosmetics, shows naked enemies of the victorious pharaoh in apparently circumcised condition. The date of the event so recorded—of the merging of Upper and Lower Egypt—is either 3100 B.C. or three centuries earlier, depending on the specialist consulted.

By the fifth century B.C. information about circumcision is a matter of record. Herodotus of Halicarnassus (who had traveled nearly everywhere worth traveling in the world of his day and heard something about the rest) reported with some perplexity that the Egyptians valued cleanliness above comeliness and so did those who had learned the habit from them. He mentions the Ethiopians in this connection but admits to some uncertainty as to which people really held priority of discovery.

The Arabs had adopted the practice by at least 342 B.C., when later Greek travel writers reported it. Some Arabian tribes were to elaborate circumcision into a supreme test of courage, one through which not all would live. Deferred until the young man was about to marry, the Arabian operation involved a partial flaying alive, the skin being peeled away from the stomach, pelvis, penis, scrotum, and inner legs. The prospective bride was invited to witness the ordeal and was privileged to refuse the match if her intended so much as flinched. Such rites were later forbidden by the Code of Islam, which nevertheless insisted, "Circumcision is an ordinance for men and honorable for women."

Retooling the nether regions, female

The feminine procedure requires, at minimum, the removal of the clitoris, and sometimes the labia as well. "If this organ is not removed," say the Dogon of modern Mali, "a woman will forever be partly male, just as the man with prepuce intact will be partly female." Surgical intervention thus resolves the sexual ambivalance which a thoughtless Nature failed to edit.

The Somali of the African Horn were less concerned with ambivalence than with defense. They sought in clitoridectomy to achieve the padlock a neglectful Nature had failed to provide. How else could a girl's virginity be safeguarded? And so it was (and maybe still is) that the little girl of six or seven underwent an operation during which portions of her external genitalia were removed. What remained was then neatly stitched with thorns until the vaginal opening was welded nearly shut. At marriage a second operation partially reopened the channel, which was again partially closed after the birth of a baby. "But," said Joy Adamson, portraitist of East African tribes, "only the first wife is made to endure this ordeal, for which I am very relieved."

Baubles, bangles, and labrets

"From a very beautiful woman without ornament, men turn away," said Marcel Griaule's Dogon informant. He added, "To be naked is to be speechless." And thus to be not quite human. Nobody wants that. Indeed, the architecture of the body practically ensures the application of ornament. Never was there a better showcase. Plenty of articulations to be well hung with baubles, bangles, and beads. But also many cartilaginous extrusions which can be perforated to accommodate a variety of other ornaments. Sometimes the philosophical rationale for the puncturings is known, sometimes not.

"Let's have a baby-blackening," say the nomadic Caingang of southeastern Brazil. "Let's make a child." This means blackening the father and puncturing the baby's lower lip.

"If we don't pierce the baby's ear," say the Bella Coola of the Northwest Pacific coast, "he won't live long."

"With pierced ears the baby will be proof against lightning," say the Arapaho.

"The girl with a pierced lip," say the head-hunting Jivaro, "is ready for marriage."

"The girl with a pierced left nostril *is* married," says the Hindu, and goes on to prove it with a nose ring of such heavy gold, so encrusted with gems, that it must be anchored to the ear with a golden chain.

"Piercing our nasal septums makes us look fierce," say the Fijians.

"Piercing our noses makes us men," say some warriors of New Guinea.

The nose knows

The gorgeously ornamental nose is popular throughout Oceania, especially in New Guinea, where it tends to be quite a prominent appendage in any case. What one has, one likes, and so large noses are much admired. Mothers often pinch and press to tease out Baby's nose to the desired shape. Art forms exalt the nose, lengthening it, so that, trunklike, it connects face to nether regions. Some scholars see in the beaky New Guinea nose yet another of a long list of so-called phallic symbols. To one observing the New Guinea sing-sing, at which young girls make their "debuts," the notion is apt to make some behavioral sense. For these parties, guests are seated in line—man-girl, man-girl. In time to the music, the girls sway first to the left, then to the right, rubbing noses with each partner in turn. As excitement mounts, couples may extend the encounters up to five glorious minutes, face against face, ornamented noses pressed quite flat.

"The nose knows!" Jimmy Durante used to crow while tapping his greatest natural resource. Not for Gê-speaking people of the central Brazilian wilderness. The sense of smell, they insist, is for animals, and who would want to exalt it with ornamentation? Gê tribesmen call attention to the nose only when certain rituals require animal impersonations, and then it is blackened to connote nonhuman associations.

None of your lip

The proper human being hears and obeys the rules of society. The proper human male sings and orates. Gê tribes of the region, therefore, concentrate on ear and lip, though the order of perforation varies and so does the type of ornament. The Tchikrin (introduced in Chapter 1) equip babies with large, cigar-shaped earplugs. At weaning time, attention shifts from ear to mouth. The little boy's lip, pierced in infancy, is then gradually enlarged. As the father of a child himself, he will be privileged to wear lip disks four inches in diameter.

The related Suya postpone the ear perforating until puberty, when heedless children must begin to mind their manners. Male lips are pierced when their owners are expected to behave like men. Ear disks are painted white for coolness and passivity; lip disks are painted fiery, aggressive red. The Suya may casually empty his pendant earlobe, anchoring the loops of skin about the ear itself. He never removes his labret, his lip ornament. Without it he can hardly call himself a man. Even at night, while he may tilt it up over his face so as to display the engravings on the underside—the patterned swirls which mimic the constellation we call the Pleiades (itself a design on the big lip disk which is the sky)—he keeps his labret firmly in place.

Anthony Seeger, ethnographer of the Suya, is among a handful of specialists interested not so much in how the body is adorned as in the meaning of adornment. He thinks it likely that lip ornaments everywhere are keyed somehow to an emphasis on the spoken word, on eloquence and song. It is a persuasive theory but hard to demonstrate worldwide, for the use of the labret is sometimes decreed for women, not for men. And while women, it is true, are responsible for developing in their children the power of speech, they are seldom permitted public oration.

The political gathering punctuated by harangues is well known, for example, to Nilotic herdsmen of East Africa. Women may be seen at such an event but are not heard. Yet one finds women's lips more frequently plugged than those of men. One female use of lip equipment is reported by Elizabeth Marshall Thomas for the Dodoth. To remove the plug and put out one's tongue through the aperture constitutes the ultimate insult. Of course, one can extrude the tongue easily because the two lower incisors are missing, both sets having been removed long since.

Nothing but the tooth

The importance of the dental gap is fast declining in East Africa, though from time to time one can still hear among African students from the area exchanges similar to the following:

HE: I thought you were going to the dentist and get those awful teeth out.

SHE: Father said I didn't have to.

HE: What! Do you want to look like a donkey?

SHE: Nobody thinks that here.

HE: They will at home. I won't marry a donkey.

SHE: Then don't!

Several authorities relate the custom, at least among the Masai, to a realization that the life of a person stricken with lockjaw can be saved by feeding him through the dental gap. Others believe that Nilotes seek to make themselves—dentally, at least—as much as possible like their precious cows.

Now the evolutionary road to humanity has been studded with dental signposts. Little by little we shrank our primate fangs, grew larger, flatter molars, and acquired a jaw redesigned for grinding. For many of the world's people, those natural changes are simply not enough. More tinkering is necessary to make the distinctions crystal clear. In much of mainland Southeast Asia and its island fringes, one finds the custom of breaking to the gum line the four upper and four lower incisors (sometimes the canines also). Whatever dangling roots and nerves remain are clipped, and the whole filed smooth. Sometimes the uppers are broken off and the lowers filed to points. And that's not all. The entire denture is then blackened completely and permanently.

"Even when the jaw is clamped shut," say the Alorese, "one should always be able to see the pretty pink tongue." A favorite Malay poem compares the mouth of the beloved to a pomegranate with its black pips embedded in the bright red fruit. Flaming color is guaranteed throughout Southeast Asia by the habit, nay the social convention, of chewing betel. This popular cud includes the fruit of the areca palm and a pellet of shell lime wrapped together in a leaf of the betel pepper plant. (Tobacco or various spices may be added.) The result for the chewer is a mild "high" and a mouth and saliva dyed a vivid red.

The fondness for black teeth ("What? White teeth? Do you want to look like a dog?") is said to be an adjunct to betel chewing. To this, opponents reply that the Burmese, who chew, don't blacken, and that the Japanese, who blacken, don't chew. Correction. The Japanese *did* blacken but do so no longer.

Archaeology does not supply a minimal antiquity for the custom in Southeast Asia, but history does, at least for its appearance among the Japanese. They are mentioned in Chinese annals of late Chou times as the "Black Teeth People." By the sixteenth century only samurai blackened, and the teeth of dead soldiers were often dyed so as to impute posthumous honors. Since 1870 the custom has been forbidden to males. White teeth for women became respectable only in 1901 with the example of the empress. Some tradition-minded geishas still hew to the standards of a bygone age.

The French anthropologist Pierre Huard thoughtfully provides the Japanese recipe for black teeth:

> Take three pints hot water and one-half cup saki.
> Add iron reddened in the fire.
> Let the mix stand three days.
> Skim off the scum that rises and place this near the fire.
> Add powdered gall nuts and iron filings.
> Reheat.
> Apply to teeth with a brush.

Throughout South America, blackened teeth are here and there in vogue. Nicole Maxwell, the American explorer and student of ethnobotany, thinks she knows why. The leaf of the yanamuco tree, which temporarily darkens some tribal teeth, also protects them from decay. Would we trade "whitest white" for emancipation from the dentist's chair? Probably not.

The patter of little feet

Only in China was the sculpted foot a beauty mark. Actually it was more than that; it was a national obsession, one which gripped the Chinese imagination, male and female, for at least a thousand years. One source credits a clubfooted empress of Shang Dynasty times with originating the custom. Others point to a court dancer of T'ang times (circa A.D. 900) whose naturally tiny feet were all the rage. Other ladies of the court, certain that anything a dancer could do they could do better, began to cram their feet into ever smaller shoes. Eventually the process was begun in youth, before a little girl reached the age of ten. The four small toes of each foot were bent and tied under the sole. And the natural thrust of growth, pushing against ever tighter figure-eight bandages, bowed the arch upward so that heel and toes nearly met. However the child cried with pain, Mother must not relent, for it was said, "A plain face is given by heaven, but a poorly bound foot is a sign of laziness." No truly loving mother, through misguided softheartedness, would condemn her daughter to a less than fortunate marriage.

The three-inch foot was thought ideal. Just the size to fit a man's palm, just small enough for a man to nibble playfully or even pretend to swallow. The tiny bound foot (the so-called golden lotus) was the sex symbol of the age and the ultimate focus of eroticism. Even the smell of putrescent flesh under the bandages—for the inhibited circulation often caused toes to rot—was considered piquant by the true lotus connoisseur. For women themselves, the feet became erogenous zones. To fan a woman's desires to fever pitch, the canny lover had only to rub her feet. There were, of course, social consequences of the bound foot, perhaps anticipated, perhaps not. The lotus lady was not one to travel much abroad—not on her own two feet in any case. She was homebound and therefore chaste. Thus were the proprieties well served by the mutilated foot in its tiny, pointed slipper.

The pointed head

The deformed foot is understandable in a society with great differences, male and female, and much attention to the symbols of prestige. The deformed head, in view of what goes on inside, is a custom harder to fathom. Yet the geographical coverage is wide indeed: Melanesia, parts of Africa, and throughout the Americas in wide swaths stretching from Alaska to Peru, from Mexico to the Antilles to the Mississippi Valley. Often pointedness, like the lotus foot, has been achieved by the use of circular bindings. Visitors to the ballyhooed King Tut exhibit saw the results of such binding in the wooden image of Tut himself as a little boy. Already his head had assumed the elegant upswept point favored by royalty in his time.

In North America the cradleboard, that ever-popular babytender, seems to have done the trick. Many native Americans, however, added yet another board in front so as to flatten the brow region. The result was a skull with a high lateral flange. In the Northwest Coast, such elevated heads were an insignia of rank. No slave child, his rounded head just as it came into the world, could possibly be mistaken for a person of high degree.

The hieratic Maya also favored the elegant pointed head for the privileged classes. But they did not stop there. The eyes of aristocratic babies were forced to cross by watching constantly a ball of pitch dangling from the head board depressing their brows. Ears were pierced for large jade spools. Ditto lips. Both nostril and septum were perforated, and, on the nose bridge, a lump of clay or putty might be modeled to produce an unindented beak, Romanesque and very much admired. (We see the extended nose plainly indicated in some sculptures of Classic times—A.D. 250–900.) The teeth were filed and bored for inlays of jade and iron pyrites. Skin was elaborately tattooed. And to complete the total picture, there was always body paint (normally red all over), a pleated loincloth, and a headdress of feathers and carvings so huge, so peacocklike, one can only wonder how it was supported. Maya art reveals a horror of the empty space. Surely this convention was begun on the Maya body.

We can see the results of drastic body sculpture on monuments and ceramics of Classic times. Living Maya described the sculptural techniques to Fray Diego de Landa, who became Bishop of Yucatán in 1579. Whether they still knew the ancient reasons for body completion and were willing to speak of them we do not know. Maya writing is yet to be deciphered.

More sculptural scenarios

A schedule of plastic effects nearly as comprehensive as those of the Maya could be seen until recently among the Toradja of the Celebes, now called Sulawesi. Anthropologists N. Adriani and Albert Kruyt describe it. The heads of both boy and girl were bound, but there the similarity ended, each sex embarking on its own program of personal completion. For boys this required:

> Penis incisions. Otherwise he might harbor bad spirits inside. Better done before puberty lest postponement endanger the rice.

Teeth shortening and blackening. Otherwise he could not be married. The married man had to serve betel to guests, and how could he chew companionably with the round, white teeth of a minor, the fangs of a dog?

Arm burning by igniting plasters of palm tinder. How otherwise could one fool the smallpox spirit?

For girls the list included:

Ear piercing. How otherwise could marriageability be announced?

Teeth shortening lest, like some ancient villainess, she bite off her husband's generative member.

Hand splitting. How otherwise would she have the power to remove objects conjured into the body of her husband?

In our eagerness to invest with significance the exotic customs of other peoples, we often fail to recognize the part played by ordinary peer pressure in maintaining ideals of the completed human form. Akiga, the northeast Nigerian tribesman who told his life story to anthropologist Rupert East, outlined the sequence of events which serve to complete the proper Tiv.

Little boys dream of having their ears pierced. The first of his playgroup to work up nerve for the operation jeers at the others until they follow suit. Next comes teeth filing. For his pains, the enterprising boy will be rewarded with compliments from the girls' playgroup: "Handsome little sharpteeth," they will chorus, meanwhile referring to the unfiled others as "crab-claws" until all get in line.

There follows circumcision and then the facial scarring that marks the Tiv. Until this final act is accomplished, the boy will be taunted day in, day out with his "silly face, like a foreigner's."

But even among the Tiv old customs change. Sometime around 1916 an enterprising Tiv youth tried out a new sort of scar, using a flattened iron nail instead of the usual instrument to achieve results. The welts were shockingly flat, and oldsters inveighed against this newfangled innovation. Mercy! It was quite against Nature! Quite against proper Tiv-hood!

But owing to European influences, girls were learning to assert their preferences in the matter of husbands, and they refused to marry men with "lumpy faces." Tribal scars were soon being made with iron nails. Indeed, many men tried surgically to flatten their old-fashioned scars and keep up with the times. And so the fashion changed.

The sculpted American

The only way our own customs differ radically from those described herein is in the rapidity of change. We have certainly entertained notions about the ideal human design and have willingly cut to suit.

Dueling scars, so admired among aristocratic Germans of an earlier time, have never been popular here. Circumcision for boy babies has, however, become, as among the ancient Egyptians, a general tenet of hygiene. The face-lift, given extra legitimacy by the wife of a former U.S. president, offers a surgical accent on youth. In a society of rapid change, age *does* wither and custom stale. So sculpt and stay young! If at all possible, stay sexy as well. That, too, is a current value of our society. If you are not sexy to begin with, get that way with surgical implants fore and aft. Nothing to it. You can succeed where Nature failed. You *can* complete the job.

We do not bind head or feet, though not too long ago we considered the tiny waist so desirable in a woman that we compressed it with corsets to a sixteen-inch ideal. Never mind what happened to internal organs along the way. Among women, ever the conservatives, the corset was not seen as an instrument of torture but rather as the armor of righteousness, the uncorseted female being a "loose woman" in more ways than one. It took male doctors to alter that custom.

We tend more and more to consider teeth erotic. Perhaps not so fervently as the Rwala Bedouin, who once wrote poems to a woman's eight small incisors and longed to be tooth-to-tooth. But we certainly go to great lengths to constrain the teeth for beauty's sake. What child has not been initiated into the mysteries of braces and retainers? And while we would never blacken teeth, we have certainly been enticed by gold, in some times more than others. Solid gold caps in strategic locations were thought to give flash to the smile, and even denture wearers had gold teeth included. Nowadays one hears of athletes who prefer the more spectacular effect of inlaid diamonds.

Perforation seems to be making a comeback in various quarters. After the invention of the screw-on or clip-on earbob, one might well have expected the American female to abandon forever the pierced ear and its attendant complications. Wrong. The pierced ear returned in the 1960s just as young women began to root for the "natural" look. In the name of naturalism, faces were scrubbed clean of paint. Hair and bosoms were unbound. But earrings for pierced ears became the essential adornment, sometimes the only permissible adornment. Girls *said* they were tired of losing one clip-on of a pair. Given the strictures against all the old-time vanities, however, that explanation hardly fills the bill. Doubtless there were deeper metaphors at work. Perhaps pierced ears were seen as a return to tribal customs, the artifices of which seemed somehow more "natural" and therefore more acceptable. Perhaps they represented a return to the habits of our own simpler past. Custom need not, after all, be rational to be compelling. And who's to say what is rational, anyhow?

Some men have submitted their ears to the needle, usually only one ear. In older times, the single-earringed male was presumed to be a man of the sea—simple sailor or pirate, perhaps. Today the single stud or tasteful ring is often an announcement of membership in the gay community. Or the mark of the motorocycle devotee.

Tattooing in our society, as in Japan, has also carried occupational overtones. Sailors and truck drivers were among those most likely to put screaming eagles on their chests and adorn a flexing bicep with MOTHER. Now the tattoo artists tell us that women are their best customers, preferring to highlight sexual charms with flowers and butterflies, strategically placed.

For at least one person the bare mention of tattoo is enough to provoke a shudder. Hidden in her kitchen cabinet is a jar holding a snippet of pickled skin on which tattooed numbers can still be seen. The skin was once her own, and so were the numbers, placed there in a Nazi death camp.

Her body and many others had been engraved, revised in conformity to a special view of the world which then prevailed. In that world, some persons were to be demoted from humanity, their life worth reckoned as zero. Marked indelibly with the badge of nothingness, they could be eliminated like unwanted cats. And the executioner need suffer no guilt, no self-reproach.

By human hands was humanity made less. By human hands expunged from humankind.

3

WALL-TO-WALL BODY

The Image Transferred

I paint myself, therefore I am. I paint into being my condition, my position, my aspiration. Therefore I am human. Therefore I am different.

I alter my body, molding it into what should be, accepting pain to make it so. I complete what Nature only began. Therefore I am human. Therefore I am different.

Thus reads the testament of adornment.

But when artists of the Eurasian Ice Age began to alter the shape of clay and bone as well as their own yielding flesh, when they began to apply paint to cave walls as well as to their own skins, it was not the human condition they portrayed, not their uniqueness they celebrated. They looked instead to the world beyond the circle of firelight, beyond the mouth of the cave. And the world they saw did not include the human presence. It was one in which the hunter moved warily, knowing himself to be a trespasser, a small being of no significance in the grand scheme of life.

It was large, that world of thirty thousand years ago, and it was cold. Brief summers were succeeded by long winters, and humankind needed snug shelter and warm clothing to survive. And nourishing meals. These were ever at hand, for it was also a world of great abundance. Plains and tundra stretched from the British Isles to the Far East. And over this vast openness roamed herds of bison, far bigger than the remnant animals of today, woolly rhinoceros and woolly mammoth, huge wild cattle and wild boar, reindeer and red deer, and horses fleet and wild. And on these herds preyed bear and wolves and felines of all sorts. Man, too, was a predator there, but one, it seems, who wished to remain invisible and anonymous.

However abundant, this landscape was not perceived merely as smorgasbord. Its denizens represented power as well as food. Mighty in size and in magic, they were rulers among whom the human hunter walked carefully and with awe. No matter what his skill, no matter the numbers of mammoths he and his fellows could trap in bogs, no matter how many horses they could drive with fire over cliffs, always there was the belief that such killings had been allowed, had been ordained. For beings so mighty could not be taken; they gave of their own flesh willingly to sustain human life. And they were moved to generosity when the hunter was mindful of the proprieties, of the rites of renewal and restitution which drew the animal soul again to earth, again to feed the children of men.

And so it was that the hunter killed cleanly and without pain, killed with religious observances, killed what he dreamed, for in dreams the great Animal Masters announced the gift. They announced as well the fate decreed for a seeker of visions, the honors he might be accorded among the power beings and his human peers. So it was that human beings patterned themselves after these Animal Masters and strove to please and be at one with them. So it was that family groups came to incorporate certain of the great beings, identifying with them and forbidding their flesh at the family meal, for one does not, after all, eat family, save in sorrow and in remembrance.

The great beings were thought capable of speech and capable of transformation when they chose. They might appear as human or as animal or as a combination and, in fanciful form, might serve as guides and mentors in human destiny. Clearly, the creative force which permeated this very personal sort of world was fluid, and ultimate reality might be very different from what mere eyes perceived.

Of myth and magic

Is this a true story? Yes and no. It is, first of all, a composite image, woven by the storyteller out of the habits and beliefs of hunting folk who still follow the old way of life. It is woven as well from descriptions of life patterns which remained intact just long enough to be studied by travelers of our own time and our own world. Most of these ways are or were followed by hunters of smaller game and with limited ranges. And their beliefs are likely to be more nearly in tune with those of Mesolithic folk, hunters of ten thousand years ago, hunters who had learned to make do with scantier resources when the plenteous Ice Age came to a close.

One must remember, too, that each people possesses its own reality, complete and separate and unique, not to be superimposed on another life way. And so we can never plumb with certainty the nature of Ice Age reality. Still and all, approximations can be drawn, and insights gained, and an approach to understanding. For there are to this day some hunters who mingle art and life, who carve and engrave, who paint on body, on earth, and on rock walls, sustaining thus the mighty balance of the universe. For them, the animals are still sentient and consenting, and the world is full of a whispered past with its great doings and its heroes who were part animal, part human, and more powerful than either one.

Of facts and artifacts

One can see this world plainly in the archaeological remains. Move back in time before the explosion of Ice Age art, and one encounters the Neanderthals. At least seven of their habitations include what can only be called sanctuaries. In them, cists, chest-like enclosures of stone, hold the skulls of cave bears bracketed and supported by long bones, often with these thrust through the empty eye sockets. Now, Neanderthalers of the European alpine region ate cave bears. Apparently they also felt the need to atone for so doing. They were not unique in this respect. Living hunters right across the Siberian wastes and into North America have also slain the bear and honored his remains.

The same ambivalence can be seen in small bone plaques carved by artists of the more recent Ice Age, perhaps fifteen thousand years ago. Often these depict ritual. One shows the realistic head and legs of a bison terminating in a fleshless vertebral column. In another we see a horse and bison realistically gushing blood. They are surrounded by human figures in attitudes of reverence. Clearly these are scenes of sacrifice. But if the sacrificial animals are superbly engraved, the hurrying devotees are not. Though rendered with the selfsame instrument held by the selfsame hand, there is a vast difference in style. Human figures are barely suggested in a few crude grooves, which reveal—in addition to reverence—nakedness, hairlessness, and sex.

On limestone walls, again the same perceptions, the same differences in style. We see a carnival of Ice Age animals, naturalistically portrayed. There are engravings of animals, bas-reliefs of animals, and paintings of animals, some in black or red outlines, some in glorious polychrome. And something more than the hunter's keen-eyed observation is revealed. Each figure is invested with the artist's admiration, his sense of its unique being, its unique impact.

With one or two exceptions, the murals do not depict "scenes." The animals are not grouped in natural herd patterns. There are no discernible elements of landscape. Often paintings appear in the deepest recesses of the caves, accessible only through the use of torches and lamps. (Occasionally, the ancient lamps are found.) Often the "canvases" chosen are in narrow niches or on ceilings to which the artist could have climbed only with difficulty. These chosen spots are painted and overpainted, engraved and overengraved, sometimes into an almost undecipherable tangle of forms. It is as though the location itself carried power, as though the act of painting, rather than the finished product, carried power, was effective in some way unknown to us and perhaps unimaginable.

And here, as in the decorated plaques and rods of bone, human faces are fleeting and elusive, *in* but not *of* the glorious life portrayed. They appear as shadowy spectators at the drama of the mighty beasts—cameramen, as it were, who now and again peer out from under their old-fashioned black cloaks and are fleetingly observed. So fleeting are the humanoids, indeed, so spare of line and form, so oddly bestial, that some scholars have called them "ghosts." At least one anthropologist has insisted they are not human representations at all—whether flesh or the spiritual essence thereof. He believes them to be cartooned weasels, otters, ferrets—all animals which, like humans, rear up on their hind feet and make clever use of forepaws.

Enter the artist, but when and how?

The human presence behind the camera is attested most concretely by the handprints that abound. And these prints, claimed the Abbé Breuil, a French cleric who devoted his life to a study of the painted caves, heralded the beginning of art. Here where men sprayed powdered ocher, or blew it from a tray, around instead of on a body part, they learned that body images could be transferred. It is an appealing notion and one which appears to find confirmation in recurrence. Rock faces the world over bear the impressions or the stencils of hands. In one Australian cave—the Tombs, Mount Moffatt Station—one can even see the stencil of a whole human body. If there were more of these, or even more evidence of the human form naturalistically rendered, one could accept such a beginning. But there are not. In Australian Aboriginal art produced today, handprints mark participation in ritual. The feeling is somewhat akin to that which prompts the pious Catholic to light a votive candle in church. The handprint in European caves appears to carry a similar meaning, for it can be found in clusters and friezes around animal figures of imposing size or elegance.

If hands can be stenciled on cave walls, the Abbé Breuil seemed to think, why could not these same hands dig fingers into the soft clay that often coats the limestone? Why could they not swoop and paddle, doodling, so to speak, on the walls, outlining shadows and, by chance, creating figures? Why not doodling as a next step in the beginning of art? Indeed, we can see among the clay meander figures of Altamira, that most spectacular of the Spanish caves, the finger-engraved head of a bull. But Altamira is late in the history of cave art. There are few early examples of this doodling, and none near cave entrances where such idle sport might well occur. No, the finger tracings, the mazes of "macaroni" marks, are as deep in the caves as the sanctuary beasts and, like the stenciled hands, offer evidence of participation. In what?

Alexander Marshak, a research associate of Harvard's Peabody Museum, believes he knows the answer. He connects the meander marks with water, often to be found in the European caves as lakes or underground rivers. Just such meanders have been found in southeast Australia in the cave of Koonalda. Because of overlying debris, these figures have been assigned a minimum age of twenty thousand years. Living Aborigines are seen to make just such snaky grooves today, drawing their fingers down soft limestone. And these markings are associated with the Rainbow Serpent and with rites to encourage rain, streams, water.

Meanders are old, old symbols. The Abbé Breuil was right in that. And they are common. But their earliest documented appearance is not on clay walls but on an ox rib. It was found by Professor François Bordes in the cave of Pech de l'Azé, the same cave that yielded early evidence of color use. As long as three hundred thousand years ago the bone was inscribed with what may have been a river sign. Indeed, it looks nearly as riverlike as the wholly realistic meander on a bone of more recent times, engraved perhaps fifteen thousand years ago. That particular bone was found, as Alexander Marshak tells us, at the site of Rochereil, near the spot where two streams do in fact join, just as the bone figure depicts.

Another look at beginnings

Like the earliest water symbols, the earliest art forms rendered in media other than strictly body ones were not painted, not doodled, but appear to have been carved in bone. We spoke earlier of the Tata plaque, that beautifully beveled oval carved from a mammoth's tooth and reddened with ocher. It is not such a big step from that Neanderthal sculpture, worn smooth with much fondling, to the Vogelherd horse carved in ivory some thirty-two thousand years ago. Graceful, sophisticated, perfectly natural, it is the first of a long line of representational figures to follow during the Ice Age. Like the Tata plaque, it was made for use. Details of this two-and-one-half-inch masterpiece had been worn quite away before it was "killed," deeply grooved with an inverted V and discarded in a south German cave.

Other figurines have been discovered in the same site: horse and deer engraved with zigzag and parallel lines, and one small humanoid, a crude and faceless female, purposefully gouged and marked with use. Such carvings of naturalistic animals and anonymous females, featureless and stylized, all breast and buttock and belly, were to continue through the years. And whether the two had different associations—the one with high ritual, the other with domestic concerns—we cannot know. Possibilities of meaning inherent in the little "goddesses" are intriguing enough to demand treatment in a separate chapter.

In Eastern Europe and Central Asia, the ivory sculptures were to constitute the whole focus of art and rite. But Western Europe with its limestone caves offered a different medium of expression. From engraving in bone to engraving on a rock wall requires, says Marshak, but another short step. The oldest of these wall engravings, deeply grooved and re-grooved, include representations of the female vulva, abstract signs as crude of execution as the fugitive human forms, and many animals. Here, perfectly remembered, perfectly lifelike, the great beasts of the open spaces are first enshrined. To their engravings yet another innovation would bring the ochers with which bone carvings had long been rubbed.

Interpreting the evidence

"Art for art's sake."
"Hunting magic."
"Rites of increase."

For over a hundred years specialists have debated the meanings of the engraved and painted caves. New theories have recently been proposed by two French scholars. Disdaining the educated guess, they have sought, rather, to learn and to classify what is actually there, organizing all according to number and placement and association with the abstract signs.

Professor André Leroi-Gourhan, having studied the contents of some sixty-five of the painted caves in France and Spain, notes that the two animals most frequently represented are bison and horse. These were not necessarily the animals most prized in terms of food, though they were eaten. Venison appears to have been favored as a menu item, if we can credit the evidence of archaeology. Bison and horse, each with subsidiary animals and with associated signs, appear to have had another significance. In the view of Professor Leroi-Gourhan, that significance was sex, human sex.

This is not to be read in genitalia. The animals are mostly neuter in this respect. And although the crude human forms are so identified, depictions of the sexual act are banned as by the hand of some unseen censor. ("Prudish," the good abbé was to label the art of the caves.) Yet Professor Leroi-Gourhan sees a sexual division, so that bison represents the female sphere of life and horse represents the male sphere. Onto the tundra with its roving herds he projects a body metaphor that is human and sexual. Freudian, some critics have called this interpretation, an imposition of modern images.

Annette Laming-Emperaire, whose close studies of Lascaux have been much admired, also sees in cave art a glorification of the pairs of opposites, but not quite the same ones identified by Professor Leroi-Gourhan. She notes, in addition to male-female, the dualities of daylight and dark, domestic and ritualistic, the art of the cave entrance as opposed to the art of the sanctuaries. And for her, horse is female and bison male. One remembers that Australian Aborigines to this day assign totemic figures to the sexes as well as to clans and cult groups, moieties, phratries, and travel bands. This may be the sort of dichotomy the French scholars have in mind.

Alexander Marshak prefers, rather, to associate all the

hooved and herd animals with females and all the carnivores with males, but he quickly notes that all grand schemes are far too simple, too all-encompassing to account for meanings which, with all their nuance and variety, must ever elude classification. Some of the meanings, he thinks, have to do with perfectly practical observations of time and season as well as with the celebration of sacred things. His use of variously powered microscopes and infrared light has revealed in the art forms more than meets the naked eye, has transformed what were long thought to be "male" darts and barbs into well-observed plants, and what were thought "female" circles into corrals, complete with schematic beasts. He finds on walls and bone plaques evidence of ritual participation, of use and reuse, and of myths oft retold.

Explaining the human image

Yet, in his findings, the human presence is no less peripheral to the animal fair, no less elusive and anonymous. Naturalistic renditions of the human form are very rare indeed. Among them one can list a frieze of female lower torsos at the cave of Angles sur l'Anglin; two faceless women in the cave of La Magdelaine, posed as if modeling for a contemporary art class; from Petrkovice in Czechoslovakia, a girlish torso carved in black hematite. Naturalistic faces are fewer still. From Brassempouy, there is the tiny ivory head of a woman, elaborately coiffured but without a mouth. From Dolni Vestonice, in Czechoslovakia, there is an equally tiny head. It was meant to be a true portrait, perhaps of the woman buried within the walls of that mammoth-hunter's encampment. The skeletal face gives evidence of partial paralysis, and so does the ivory head. In life the woman may have been a shamaness, given power by virtue of her disfigurement. Obviously, sculptors of the Ice Age had the ability to capture the human likeness. Whether they had the wish or the will to do so is something else again.

In the French cave of La Marche, quantities of engraved stones were found, many of them bearing, in addition to mammoths and other beasts, tangles of human forms and faces. Dr. Léon Pales and his associates at the Musée de l'Homme have resurrected some of these. As in all the other caves, the figures are mostly naked (odd for such cold climes) and without facial or body hair. None displays a weapon or appears to threaten the animals. As in the other caves, crudeness of execution is the order of the day. (The Abbé Breuil thought the crudity was intentional, designed to foil the sorcerer's evil magic.)

Some of the faces at La Marche are snub-nosed and pop-eyed. Others carry snouts or beaks. One of these was apparently intended to be a serious portrait of a young man in profile. The handsome neck, the tip-tilt of the nose, the almond eyes are real and attractive enough. But the face is pulled forward, muzzlelike, in curious evocation of the graceful stags Ice Age artists presented so faithfully.

Everywhere in the caves, on stony slabs, on decorated plaques, cartoon faces bear the indelible mark of the beast. And whether such marks were meant to represent masks or whether they describe what the artist believed he saw or wanted to see, we cannot know. Everywhere the animal-headed, bird-headed figures are shown in a curious posture, body inclined forward, knees bent in a semi-squat. Some scholars call this a dance posture. Others have seen in it an animal imitation. Still others, more interested in line and form than in meaning, have called attention to the ease with which such a stance can be transformed into the curving back of bison or mammoth. So often are cartooned females shown in this posture that a few scholars have inferred a Paleolithic preference for the ventral-dorsal position in the sex act and interpret the little drawings as stylized come-ons—pinups, so to speak, of the Ice Age. Coal-carved pendants of the squat position were worn in some areas as talismans. And so pervasive was their meaning that we find necklaces of such pendant figures millennia afterward in the Neolithic city of Çatal Hüyük.

If one compares such tokens to those of praying hands, worn as pendants by pious Catholics today, one might find yet another plausible interpretation of the "Paleolithic squat," reading in it an attitude of reverence. Certainly it is true that males as well as females are seen assuming this posture. Arms upraised, they bend the knee before bison, horses, bears, and huge fish. Indeed, there is at least one known sculpture of lifted arms with head and upper torso. Carved in reindeer antler, the little figure was found in the cave of Laugerie Basse. Missing, apparently broken off, are the lower torso and legs, which may once have presented the complete attitude of piety. That it was a traditional stance is reinforced by geographical coverage. We can see such postures, assumed by engraved humanoid figures before lowering beasts, on the rock walls of the southern Atlas Mountains, in the Sahara, and on cliffs in South Africa. Apparently it took quite some time to bring the human hands together in prayer and the knees fully to the ground.

The animal alter ego

As for the mark of the beast, there is yet more to say. Some of the figures do not merely evoke the animal image. They wear undoubted bison heads and pelts and carry bison hooves. They are composites—antlered, owl-eyed, bison-bearded, wolf-tailed, bear-pawed—and only the feet to give away the human underneath it all. Such figures drawn by living hunter-artists surely bear many meanings. They may represent the shaman, the man of knowledge, with spirit connections, with animal familiars and guides. They may represent a creation myth with animals turning into human beings, and also the other way around. They may represent elaborate metaphors and conditions—animal attributes, so to speak—a sort of pictorial "strong as a bull" or "stubborn as a mule." They may call on the feral alter ego. There are still people in the world who believe that for every human soul there is an animal counterpart and that the death of one will cause the other immediately to give up the ghost.

A few of the plaques depict pregnant, hairy women, lying prone beneath or near large male bisons. In a jumble of wall-painted figures at Trois Frères, an upright bison-man can be seen to follow a retreating reindeer with human forearms and a female bison, genitalia enlarged, her head turned to watch the strange being approaching. Perhaps such figures reflect the belief—pervasive through time—that the power beings not only feed humankind with their own flesh but, now and again, consent to sacred marriage. Diamond Jenness, who lived among the Chipewyan, Athapaskan-speaking hunters of the Canadian

Northwest, heard such sentiments many times. Said one of his informants:

> We know what the animals do, what are the needs of the beavers, the bear, the salmon, and other creatures because long ago men married them and acquired this knowledge from their animal wives. Today the priests say we lie, but we know better. The white man has been only a short time in this country and knows very little about the animals. We have been here thousands of years and were taught long ago by the animals themselves. The white man writes everything down in a book so that it will not be forgotten. But our ancestors married the animals, learned all their ways, and passed on the knowledge from one generation to another.

Whether as mythologized mate, awesome power, sacrifice, or atonement, it was the animal image that dominated the art of the Ice Age. The hesitant, scurrying, awestruck humans who shared the stage saw themselves as minor characters, extras and supernumeraries whose poverty of costume and presence would be noticed little if at all. Only in animal guise, wearing animal masks and animal skins, did they acquire identity and move briefly into the spotlight.

That, at least, is what their own self-representations tell us, their very-far-distant descendants. Doubtless they strove among themselves for influence and rank. Doubtless they prepared their young for the responsibilities of age, comforted the woman in labor, gave solace to the dying. Certainly they strove to keep body and soul together, wandering far afield in search of plants and game. But the look of these strictly human events is nowhere recorded. There are only the mute and abstract signs, the pictographs which may speak of family ownership of certain rites. They may also document the participation of family members in these rites in much the same way that Southwestern Pueblo people, on pilgrimage to salt sources, leave behind on the rocks engraved emblems of their several clans. Indeed, the signs may carry a hundred different sorts of messages—glorious, wholly mundane, even wholly trivial. Many of the pictographic statements made by Pueblo people during the last few hundred years are trivial. Anthropologist Mischa Titiev translates one—a collection including a stylized tree, a rather realistic pistol, a stylized vulva, and a coyote head—as follows: "Here a Bakap clansman had intercourse with a Coyote clanswoman."

The Ice Age signs, alas, cannot be translated. They can only tantalize and suggest human happenings. They are there. So are the beasts in whose images human need apparently found both location and transcendence.

Enter a new age

Sometime around twelve thousand years ago, the scenery changed and the drama was rewritten. Glaciers of the far north and of mountain ranges beat a permanent retreat. Seawaters long trapped in ice devoured old shorelines and crept inland. What had been tundra sprouted forests of deciduous trees. The great herds followed the ice, became rare, became extinct. Other species of animals flourished; the cast of characters changed. And the hunters, waiting in the wings, vied for the foreground.

In these new times, the Mesolithic or Middle Stone Age, the enigmatic signs, the pictographs of old, were painted on pebbles. Impossible to know whether the ancient meanings continued to speak or whether their use represented an antiquarian exercise. Mesolithic art was not produced in the depths of caves but on cliff faces and under rock shelters. And though certain locations—clefts and gorges difficult of access, unused as habitat sites—appear to have enjoyed more preference than others, it was essentially an art of the open air. In most places, the emphasis was on painting. There are areas of France and southern Spain, of North and South Africa, of Australia, where we can see the vivid one-color line drawings of the Mesolithic obscuring the deeply engraved figures of earlier times.

Mesolithic hunters took delight in their own doings. We see them gathering honey (a painting complete with angry bees). We see them conducting complex rituals, including the sacrifice of animals and possibly of men. And because numbers had grown, because in those warmer climes people had traveled, imposing their strangeness on one another, we see them at war—group against group in equal strength and equal advantage. We see their weapons, for so long banished from the art of the caves, and we see those weapons turned against the animals as against one another. Human figures and animal figures are rendered in parity, equivalent in energy and in cunning. There are few beings partly man and partly beast. Often when animal attributes appear, they are clearly an aspect of costume, to be used as often in hunting disguises as in ritual. The powerful beings of another age, the animal masters and animal mates, have not been altogether displaced. (We know by way of still-living myth that such beings would long continue to influence the affairs of humankind.) It is just that the human form and human events have gained in stature. The dailiness of human life has come to be worthy of memory and of record.

Much the same sequence can be read in the Atlas Mountains and the rocky massifs of the African Sahara. Ancient deep engravings of great wild animals, of animal-headed men, and of men in postures of reverence give way to the painted stick figures of "Little Devils," the hunters as they saw themselves. Over these images are drawn and scratched hunters of more substantial build, surmounted by the faceless, decorated globes which prompted Henri Lhote, the explorer and Saharanist, to call these figures "Martians in space helmets."

Both animal and human forms were inflated in size beyond all natural proportion. In rock shelters big enough to suit the graphic ambition of the age, artists painted what can only be described as gods. One of these is often called the Great Master of Sefar. Though not entirely naturalistic, not drawn with felicity and grace, he is nevertheless awesome and imposing. Pregnant, fishbowl-headed women surround him, their arms uplifted in supplication. Some are lying supine in poses reminiscent of certain figures carved in Ice Age ivories. Though the Great Master wears a tail and a mask crowned with antelope horns, he is no were-antelope, assuming feral guise. Nor yet is he a mere shaman bearing animal attributes. He is a thoroughly human projection of the sacred and the powerful.

A little way within the gloom a roebuck raised his eyes,
Brimful of starlight, and he said, "The stamper of the skies,
He is a gentle roebuck, for how else, I pray, could he
Conceive a thing so sad and soft, a gentle thing like me?"

So the Irish poet W. B. Yeats supposed that living things would frame, each in his own image, the powers of the universe. Yet humankind had not done so until now, here in a rock shelter, in the great grassland that was to wither one day into the Sahara Desert.

And yet another reality

Sometime in the seventh millennium before Christ, herdsmen entered that grassland. Whether they had traveled great distances with their domesticated cattle, sheep, and goats, or whether they had somehow tamed the wild forms indigenous to the region, we do not know. Certain it is that these newcomers, Neolithic folk who had learned to husband and conserve their source of sustenance, who had learned to dispose of it as seemed best to them, to harvest it at need, saw the world through different eyes and with expectations vastly different from those of the hunters who had preceded them.

They were painters who decorated the faces of rocky cliffs. Their art bears echoes of the exquisite naturalism of the Ice Age. But it is much extended in subject matter, finding its best expression in the human form. If the native antelope are elegant, if the cattle herds are rendered with meticulous care, so too are the hunters, the herders, the women gatherers, the mothers, and the young. No event was too small, it seems, too commonplace for the artist's eye and the artist's hand. The rocky walls unfold a whole panorama of life among the "Bovidians." We see people of different racial groups mingling here in apparent harmony. We see drawn in painstaking detail the different hair colors and coiffures of ladies of fashion who ride haughtily on oxen back. We see what can only be scenes of judgment, of battle, of ritual. Not, however, very much of that; the Bovidians strike us as a thoroughly secular folk. Some of their cattle do indeed carry marks of distinction—what we might think of as "halo images"—between their horns. Perhaps there obtained in this region a sense of mystic communion between herdsmen and herds. Much the same sort of feeling exists to this day among pastoral peoples of the Upper Nile, where the Bovidians doubtless retreated after the great drying-up began. But men there know themselves to be masters of their herds, make no mistake about it. And so did their cultural forebears, who had no qualms about culling their animals and butchering them. The paintings tell us so.

Some of the human figures painted by Bovidian artists bear animal attributes. These appear somehow to be rather statements of position than of composite being. As men of rank among the ancient Maya wore tunics and gloves of jaguar hide (their latter-day descendants, remote from the modern world, wear them still); as the king of the Swazi Nation, independent yet enclosed in the Republic of South Africa, wears in public a leopard tunic and a mantle of oxtails and is addressed as "Our Bull"; so in Saharan paintings the mark of the beast may well have been the signature of rank. Three painted females, bird-headed and elegant, remind us of Egyptian deities whose special powers and concerns were conveyed by animal metaphors. Some scholars like to locate in the Sahara a cradle of art themes and forms both for Dynastic Egypt and points south. And radiometric dates do indeed suggest that the pioneer herders and their art preceded the earliest Egyptian farm communities by at least a thousand years, and dynastic art by two thousand more.

The human image triumphant

Wherever the knowledge of domestication spread, human numbers multiplied. And wherever these coalesced into settlement, there arose the complex social conventions and beliefs that were to leave their own imprint on the human image. Some of this can be seen in the Neolithic city of Çatal Hüyük, south of the Konya Mountains in modern Turkey. Barely an acre of this thirty-acre mound was excavated by archaeologist James Mellaart before the site was closed by government decree. The earliest vertical level yields a date of 6500 B.C., and there is unexplored evidence of much earlier occupation. Fortunately, the parts exposed appear to have constituted the ritual center of the city, built much after the fashion of the pueblos of our American Southwest, with doors in the roof of each stepped-back apartment, and entry by way of ladder.

There is much about the shrine art to remind us of the Ice Age. Wall paintings and engravings; bas-reliefs of rams, bulls, stags, and now and then the leopard. There are modeled heads of bulls and rams, including their natural horn cores. There are friezes of human hands and the ancient zigzag designs. There are female attributes—not engraved vulvas but sculpted breasts, each of which encloses the jaw of a wild boar or the beak of a vulture (to whose tender mercies the bodies of Çatal Hüyük's dead were apparently consigned for excarnation).

Indeed, there is more to the female presence here than a few isolated portions of the anatomy, more than a few figurines, fertile and fat. The Mother is literally everywhere, and everywhere in domination of the beast. As undoubted goddess, she is sometimes seen in association with a male god, whose virile attributes the horned heads may represent. Sometimes she is shown with leopards, nursing them, riding on their backs. One lovely statuette shows her in seated posture, supported by two felines, giving birth to what appears to be a human baby. It is necessary to note the humanness. Elsewhere in the shrines, the goddess is shown giving birth to rams and bulls.

Whatever the symbolic meanings at play, clearly the creative force on which all life depends is ultimately a human one. Clearly the human image is in charge, is in the ascendant as the human actor has taken charge of the drama of life. Even the great beasts carved or painted on the plaster walls are bereft of all terror and cast in a human frame. Hunters, clad in what appear to be leopard-skin tutus, dance lightheartedly around bulls and stags and boars. They chat with one another as they loose their flights of arrows. The scenes include large human casts, each member individualized in terms of activity, involvement, and appearance.

And as for the beasts themselves, much exaggerated in proportions so as to magnify the hunter's skill, there is to be seen a certain slapdash clumsiness of artistic execution. Often the hooves are painted in curious resemblance to human hands. And the giant and fearsome vultures are equipped with fingers rather than claws. One remembers how children in our own time draw cats and dogs with human smiles and human limbs. The artists of Çatal Hüyük, having learned to paint in a world ruled by humankind, tended to translate what they saw in human terms. They had grown quite out of touch with other, older realities—realities which, for artists of the Ice Age, were the only ones worth the effort and worth the paint.

And so it has remained through the great civilizations that have followed, one after another like pearls on a string. Whatever the bucolic yearnings of poets for a simpler time when man and nature were at one, humankind has remained the measure of ultimate reality.

Images yesterday and today

I look out from myself. I am a mirror, a camera, a recording tool. I reflect the master beasts.

I look at myself and see an equal to the others, my animal peers.

I look at myself and see what I'd like to be—bigger than life and twice as handsome. And the gods who bestride my universe each wear my face.

I look inside myself and see the grotesqueries, the hybrid monsters, the power beings of the mind.

In recent years the human image has again undergone reduction. First we learned that earth was not the center of the universe and that the godhead had other business abroad. Then, with an understanding of human origins, we reassumed the mark of the beast, finding in it this time not honor but degradation. Finally we lost the appeal to reason. We began to see ourselves impelled by forces and motivations irresistible but dimly understood, inner forces more threatening than all the outward terrors of the world. Retaining mastery of things and machines, we abdicated mastery of ourselves.

Like the "X-ray" designs of Australian Aborigines, designs which plumb the hidden workings of a beast, the art forms of our day are internal compositions. They interpret not so much what is seen as what is felt. The new image of humankind is compounded of private visions, of countless symbol sets known only to the artist. And the image moves through interior landscapes as circumscribed and enclosed as the painted caves of the Ice Age.

And what of the future when at last we move into the tundra of the night? Shall we see ourselves puny and small among the stars, as once we did when great beasts were many and humankind few?

Or shall we project upon the infinite a human face, shaping larger realities by what is knowable and known? For it is in the image that we conquer and, in the image, rule.

4

SAVING FACE
Masks and Meanings

I am what I do. And what I do is written in the mask I wear. I am the mask.

Every day of our lives the masks go off and on, donned, discarded, exchanged, as we move from obligation to obligation and from friend to friend. Never mind that the masks are invisible, being the facial expressions, the stances, the vocabulary, attitudes, even the tone of voice appropriate to each position, each condition of life. We wear them all the same.

"Good morning, Dr. Cureall," we say, pleating our faces into the proper "patient" folds while he assumes his professional mien, purring confidence and concern.

"Two hours of preparation should precede each class meeting. That is the rule of thumb," lectures the professor, radiating stern righteousness for erring student, suitably downcast.

Parent-child, lover-lover, husband-wife, boss-secretary, foreman-laborer: our days are studded with such mini-dramas, their dialogue long since learned by heart. The oft-repeated exhortation "Why can't you be yourself?" was and is the theme song of adolescence. But who knows, who ever knew, exactly what that was? We can hardly think of ourselves as beings separate from our social identities. For whatever is uniquely ours—personality, inner self—is inevitably shaped by the parts we play, by the invisible masks. The very word for "person" derives from *persona*, the Latin word for mask, and through it to the Greek *prosopon*, which is both mask and face.

> All the world's a stage [wrote Shakespeare]
> And all the men and women merely players.
> They have their exits and their entrances;
> And one man in his time plays many parts . . .

Of the changing nature of role and of the mask, none knows better than the actor who performs not merely the varied scripts of his own life drama but those of other men as well. It has been said that Sir Laurence Olivier, the great English actor, achieved his sensitive characterizations from the outside in, from mask to soul. His Mahdi in the film *Khartoum*, for example, came by way of dark-painted skin and red mouth, by way of stance consciously assumed, by the stride of legs enveloped in long robes, by the tense gestures of the religious zealot, by exotically accented English. These shaped the pattern of thought, directed the actor's search for motivation and the inner man.

Visible otherness

In older times such a search need not have been made, for role was vested in a visible mask, an art form in its own right. In the archaic No drama of Japan, the mask is still the message, as it is in traditional Korean satires. For these, the masks anciently worn by shamans to frighten demons were turned to social and political protest, a way of mocking the lascivious monk, the exorbitant lord. And, as protest, the drama still functions to this day.

The actors of Classic Greek drama were masked with at least

one practical end in view. Small megaphones were concealed inside the capacious wooden mouths, the better to carry the author's words to the topmost tier of the sunlit amphitheater. Yet more than amplification was intended. The mask conferred identity and conferred size, conferred heroism beyond that of the merely mortal, permitting the merely mortal actor to transcend himself for good or ill.

Otherness. This is the eternal meaning of the mask. It is the one we see carved into bone plaques and on cave walls of the Ice Age. Whether the feral features which mark fugitive human forms are intended to depict transformation into something other than human, whether they are enhanced imitation or enhanced impersonation of the great beasts, the goal is still the same: to transcend human reality and ennoble it. Some figures indicate costume: a Lascaux shaman in trance wearing a beak identical to that of the bird effigy on his staff; three little figures from a plaque in Teyat wearing bison heads and pelts down to their waists; that "bison man" from Trois Frères following a female who is all beast. Their respective sizes show too much disparity to give the man were-bison status. Clearly he is masked.

The mask true to life

Looking at this figure, traced by the devoted Abbé Breuil, one cannot but remember the paintings of George Catlin, recorder of the American Indian in the early 1800s. While among the Mandan, whose earth-lodge villages were located along the upper Missouri River, Catlin witnessed the powerful O-kee-pa ritual which celebrated Mandan beginnings, their source of plenty, and the tribal future vested in their brave young men. Thanksgiving and supplication were directed toward the buffalo, represented by eight male dancers. Each of these was elaborately painted in red, black, and white, with concentric circles marking each joint and on each abdomen the face of a human baby. Over the paint was thrown the pelt of a buffalo, horns and facial skin intact. The dancers peered out through empty buffalo eye sockets. Other animals, less prominent in the ritual, were also impersonated by painted men: bald eagles, antelope, swans, rattlesnakes, beavers, vultures. Wolf and bear dancers, like the buffalo men, wore actual hides and heads. How all these were padded and propped, their facial parts stiffened into lifelike muzzles, Catlin does not say.

Archaeology reveals one method of mask support practiced twenty-five hundred years or so ago in Kentucky, Ohio, and as far east as New Jersey. Among Adena Mound Builders there, men privileged to wear the masks of wolves and pumas submitted to the removal of the upper front incisors. Into this aperture was fitted the animal palate, thinned and narrowed, but including the foreteeth and fangs. Since this denture protruded several inches beyond the human face, it must have served to position the mask and lend it "bite." That the demands of ritual were not without a price can be seen in a male skull from the Kraft site in New Jersey. A secondary infection, resulting no doubt from evulsion, had eaten away the upper gum to the nose line before healing at last.

Once, during a Copper Eskimo séance, Diamond Jenness saw the presiding female shaman "become" a wolf with a few bone-chilling howls and a couple of wolf fangs surreptitiously slipped into her mouth. Her mimicry of the beast was perfect, and every person present (except Jenness himself) later swore she had indeed turned into a wolf and back again. So believing makes reality for us all, with or without psychotropic assistance. But imagine what an adjunct to belief the total costume can provide. And with a muzzle, with jaws that realistically gape and a realistically full display of lupine dentures, what a triumph the enterprising shaman could achieve. If, as the saying goes, women will suffer much for beauty, certainly the Adena shamans were equally willing to suffer for their art.

The mask in prehistory

Might shamans of the Ice Age have practiced such self-mutilation in the interests of verisimilitude? Archaeology has yet to uncover the evidence. But the importance of the mask, begun in those early times, continued, grew, and multiplied its meanings. In the exuberant outdoor art of Mesolithic times, we see masked men presiding over ritual, but we can also see masked hunters in pursuit of game, their impersonations serving economic rather than religious ends. Rarely in the art of those times does the human face appear as an art motif, separate, solitary, stylized as in a mask. Not until the advent of the Neolithic do masks appear to take on even minimal human attributes.

Farming skills and ways came to Europe, most specialists now believe, sometime around 7000 B.C., filtering northward from the peninsula that is now Turkey. Among the remains of ancient farm villages in Greece and the Balkans, there have been found countless figurines of stone and clay. And these figurines are all shown wearing masks, triangular or oval in shape, with bulging, slitted eyes. The mask alone, on pottery and other objects, provided an important decorative theme. Dr. Marija Gimbutas, professor of European archaeology at the University of California, believes the masks—humanoid but with horns and sometimes beaks—represent deities to be activated by living men. For over two thousand years this concept and this image of divinity continued to stir imagination and awe. Several curious forms found by Dr. Gimbutas were composites: sloping clay pylons onto which were slipped removable ceramic masks. The changing nature of the god, perhaps, or at least the changing content of human need.

The mask of death

In the same general area, among the remains of a still more recent folk, very rich in gold, Dr. Gimbutas found yet another meaning for the mask. Her excavations were concentrated on a sixty-five-hundred-year-old cemetery near the Black Sea. Grave goods of the most elegant sort accompanied the bones. But some graves, equally lavish, were without bones, holding instead what Dr. Gimbutas calls "the mask of the goddess." It is uniformly the same: life-sized and of clay, adorned with a gold diadem and with eyes and mouth represented by disks or strips of gold. The clay ears, well defined, are pierced all around the rims by gold rings, and to the ceramic chin are attached ring-shaped pendants with "eyes." Perhaps, says the archaeologist, such graves are a ritual expression of death and transfiguration or of death and resurrection—an expected new birth of the goddess. One might think as well of the cenotaph, a symbolic grave for the man lost at sea or on a dangerous venture, returned home in an equally symbolic way and laid to rest with the image of the goddess in the hope of life eternal.

More familiar to us is the grave mask bearing some sense of individual personhood. Death portraits are known to every part of the world and are found in every social context, some sophisticated, some simple. One can see in museums the masks of clay and stone, encrusted often with precious jade, which covered skeletal faces and cremation bundles in ancient Mexico. One can see the masks of gold foil which identified mummy bundles in Peru. All of these were shaped to a predetermined form, more artistic convention than individual portrait. True individuality marks the gold-foil death masks of Greece fashioned some thirty-five hundred years ago to cover the faces of Mycenaean heroes. Finding one grim and bearded visage, the excavator Heinrich Schliemann was moved to exclaim, "I have seen the face of Agamemnon!" Not so. The hero of the mask had lived at least two hundred years before the Trojan War.

First and last causes, masked

But masks are for the living as for the dead, for secular as for sacred things, and for the countless turns of event which complicate the lives of social beings. Already in Neolithic times this was so. A people of the once-green Sahara, people who may have lived concurrently with the Bovidians but whose manner of life was of a very different order, made of the stylized human face an important focus of art. The lines of the mask are sere and stern. Like the Great Master of Sefar, each is equipped with antelope horns. One of the masks adorns the body of a man, scarred of skin, bent of knee, and sprouting blossoms from shoulders and thighs. Its virtually identical descendants can still be seen today among the Senufo people of the northwest Ivory Coast. Attached to their great wooden masks are little bent legs which are said to link them to earth but which vividly recall the figure behind the Saharan mask and suggest the assimilation of the wearer by what he wore.

Throughout this region today, masks both represent and provide vessels for the Great Ones who have gone on. The Dogon, long ago driven into the high, rocky hills overlooking the Niger, call out the masks, worn by members of a special cult, for the funerals of their important men. They seek thus to contain the spiritual forces released by death. The greatest of all masks recalls the first dead person in the Dogon world, the ancestral human being. Once every sixty years it is refreshed, carved anew from a tree dripping with red sap. Activated by the sacrifice of a dog and a hen, it is carried in solemn procession through all the villages of the Dogon, summoning their people to rededication to sacred first principles and to their perpetuation through eternity.

The masks appear also at funerals of the Bambara, but they represent first principles of the graded age fraternities and announce that the deceased continues in bona fide membership and in perfect protection from grave throughout eternity. The Bambara, who live southward of the Dogon but still in modern Mali, have been attached to several ancient kingdoms now defunct. According to Pascal James Imperato (who was there), they still hold to traditional beliefs, but perhaps not for long. The power of Islam waxes steadily in the region, sweeping away representational art wherever it exists. For the Bambara, it is the art of the mask and the headdress which are under threat, for it is in the mask that belief is made manifest.

The beginning of humankind, the gift of agriculture, the continuity of life and love—all are stories told by the masks. Several types are owned by a particular age society, through each one of which an individual moves by virtue of accumulating birthdays, living the knowledge appropriate to the group as he wears the appropriate mask. Only one of these takes a human form. Mouthless, made of cowrie shells and crowned with animal horns, it signifies both skeleton and being of the first man, primal man, beautiful but without speech. Roan Antelope, partly animal and partly man, is the Bambara culture hero, bringer of domesticated plants. The other masks represent animals whose attributes, good and bad, mimic those of the human personality.

It is at the great agricultural festivals that the lessons of the animals are told:

Be like the lion; he is dignified.
Be like the hornbill; he is wise.
Don't be like the baboon; he is lascivious.
Don't be like the buffalo; he is cruel.

Masks of judgment

In other days, some of the age societies, under cover of their masks, were said to punish offenses of certain kinds: adultery, theft, and murder by poison—which may also mean murder by sorcery. Still farther southward along the West African coast, the masks assume an even more explicit role in judicial procedure. Common to this region—throughout the many tribes of Sierra Leone, Ivory Coast, Liberia, parts of Ghana, and Portuguese Guinea—is the society not of age but of secrecy. There is one for women and one for men. The existence of the societies is not secret; it is their lore which is secret and the sources of hidden power. These derive intact from "the ancestors," spiritual beings, forest beings, anciently acquired by humankind through alliance or by descent. The great unseen ones are resident in the masks, masks which empower their wearers to give judgment in secret, and in secret to order punishment, even death. So feared, so dread are the "mother" masks, commanders of their masked lieutenants, that, like the medieval clergy, they can level an interdict, can quarantine the neighborhood harboring a criminal until his own relatives dispose of him. Much of the awesome impact of the mask depends on the artist's hand, his rendering of features that are at once human and more than human, more majestic, more terrible. The slitted, dreaming eyes of the mask, the cold, spare lines, the frightful symmetry of facial plane inspire respect even in the outsider, the skeptical observer from the West.

Nobody ever knows who wears the mother masks, though there is much private guessing. He may or may not be the tribal chief, giving solemn judgment in the ordinary, everyday court of the sort well known to our world. Behind and underneath all these trappings of civil procedure moves the real power, secret, unseen, religious. It is a government within a government, engineering the election of chiefs and binding them to its decrees. The nearest one can come to an analogy is the Ku Klux Klan of our American South just before World War I. Through terror and coercion the Klan then held the power of decision. And who knew the real identity that lay behind the white and hooded forms?

The mask of kings

The power of the secret society was long ago co-opted in Nigeria and Ghana when kings appeared, sacred in their persons, vested with supernatural sanction quite beyond that of the mere forest spirit or local ancestor. Secret societies waned then in importance and lost their license to punish except at the express order of the king. Yet the masks remained as dynastic emblems, conferring on the living man the right to rule. For the Yoruba of Nigeria, the human mask is worked in beads on front and back of the royal crown. Beads hang from it in thick fringes to hide from public view the face of the living man. On the conical peak of the crown there cluster beaded birds, symbol of communication with the spirits on high. But they carry as well a more sinister connotation. They speak of the powers of evil which the living king must in his person reconcile to the powers of good in order to strike that delicate balance essential to his people's life and happiness. The king may wear the crown but looks with peril into its secret interior. He may of course be made to look there by his chiefs if they deem him unfit for office, and then is his death at hand. Thus do Yoruba restrain the tyrant by his own crown, by the mask of kings which is at once badge and code of office.

Spirits, cures, and realities

The many meanings of the mask can be seen in the New World, from one continental corner to the next, from north to south, among hunting tribes and great kingdoms, and at all the social way stations in between.

On the North Pacific coast around Canada's Vancouver Island, the Kwakiutl, rich in an area of great natural abundance, explored through masks the protean nature of reality. Observe their raven's mask, realistic as to beak which shuts and opens at the wearer's whim. But wait. Is it raven or something else? The beak splits in twain, the halves part vertically to reveal the humanoid features underneath. The soul in the flesh, appearance and its betrayal, mask within mask. For, say the Kwakiutl, each family began with an ancestor who, arriving on this earth, removed his animal mask and was human. Still masked are the speaking animals, the magical animals that present themselves for solemn pronouncements. The sky is the greatest mask of all, behind which lurks the ultimate hidden reality. And this no man may see.

The agricultural and warlike Iroquois of upper New York State dispensed with animal representations and concentrated on human features made as ugly, as frightful as possible. For these were the faces of evil and the proof against evil, faces to forbid the entrance of sickness and death and, once among the tribes, to frighten such horrors away. The masks were made in remembrance of faces, disembodied and powerful, which had appeared in the forest to solitary hunters. Whoever dreamed of the faces immediately sought ordination in their company and made for himself a mask carved first into a living tree so as to carry ever after that extra vitality. While he wore the mask he was filled with the power to heal, able to handle live coals without fear, without burns. When he dreamed of separation from the faces, he was immediately acquitted of his responsibilities.

Westward among the mesa-top pueblos of Arizona and New Mexico, the masks are humanoid but without human individuality. They are all-purpose leather cylinders with peglike wooden noses, the whole paintable to many role specifications. They represent the Kachinas, august dead who return to their homes and villages for six months of every year, bringing rain and other blessings to the living. When a Hopi dies, his dead face is covered with a white cloth mask and, according to Dr. Titiev, he is addressed thus: "Now you are no longer a Hopi. You are a Kachina. You are cloud."

But which cloud and which Kachina is a matter of supreme indifference. Like the pebble tossed into a pond, individuality is in death lost in the collective and anonymous sacredness of the group, the ultimate self-effacement pueblo folk strive for constantly in life but somehow never quite achieve.

Gods, graves, and dominion

The greatest complexity in New World masks and meanings is to be found in Mexico. In the state of Guerrero, villagers still make masks of animal and human form and wear them in festivals. All highly appropriate since, according to Gillett G. Griffin of Princeton, Guerrero was anciently the major workshop of mask manufacture for the successive great states of the region.

Oldest was that of the Olmec, who flourished more than three thousand years ago. Olmec, a name coined by later-appearing Aztecs, was meant to describe people they knew on the low-lying Caribbean coast, homeland of the rubber tree. What the ancient inhabitants called themselves we do not know, but the Olmec "look" remains. We see it in the heads they carved from gigantic blocks of basalt, some of them nine feet tall and many tons in weight. We see it in numerous altarpieces and figurines. The look: eyes at a slant; lips full, downturned, and half-open, revealing a babyish row of teeth; facial planes sensuously smooth and wholly symmetrical. Nowhere are these features more superbly displayed than on masks of jade. Some are life-size, with the round, open eyes that suggest they were worn. Some are small, made to adorn a headband, perhaps, or to be worn pendantlike around the neck. Perhaps they evoked the identity of a tutelary deity, a guardian. We know only that such masks are engraved inside and out with the were-jaguar-baby god whose stylized features mark the face of every other Olmec deity and every Olmec soul. This was the ultimate reality, the ultimate mask, this strange, exotic being, part crying infant, part snarling feline. This was the ultimate matrix within which Olmec art took form. It was more than that. The were-jaguar visage was also emblem of office and emblem of rank.

On a rock face at Chalcatzingo in the state of Morelos, Olmec artists engraved a curious "scene." We see three figures wearing were-jaguar masks. Two wield clubs to menace yet another figure. He is seated on the ground, his hands bound behind him. His mask has been knocked away and hangs ludicrously from the back of his head. Unlike the others in their elegant loincloths, he is naked, a condition which would always be a mark of degradation in Mesoamerican cultures. Is he a defrocked priest, perhaps? Or a political personage under arrest?

Among the Classic Maya, Olmec inheritors and descendants, the mask continued in uses which appear to be both religious and secular, and who knows where the one gives way to the other? In the elegant, convoluted mazes of Maya style, in the stelae, on the painted pots, we see figures of the mighty adorned with the jaguar skin and sometimes the paws that announced high rank. They wear as well the masks of deities, often with animal attributes. In one stela we see such a personage in profile, his mask in cutaway view attached to his elaborate headgear.

A Maya vase owned by the Art Institute of Chicago depicts the mighty—judges, rulers, gods—wearing human masks. And since they, too, are shown in cutaway, we can see that the profiles of the masks mimic the profiles of their wearers even to goatees, mustaches, and set-in teeth. Human masks of any kind are rare in Maya art, and portrait masks worn by living men are rarer still. Perhaps the scene represents judgment and sacrifice, as the describer, Anne Paul, believes. Perhaps it is yet another episode in the continuing saga of the Great Lords of the Underworld, the kind of lore Michael D. Coe, the eminent Mesoamericanist, reads in all funerary pottery of the Maya.

Aztecs and their contemporaries made similar use of masks. Paintings of Aztec gods, as they appear in the codices—the folded "books" of pre- and post-Conquest times, are often paintings of human faces, perhaps priests, wearing the masks and attributes of gods. We can see the painted human nose behind the fantastic superstructure of Quetzalcoatl in his Winds aspect. We see the priest as Xipe, Our Lord the Flayed One, wear the skin of a sacrificed victim cut from his living body to represent the new green of spring.

The masked mouth

Through all the successive cultures of Mesoamerica, artists have often taken care to show the mask for what it is: the means by which the wearer obliterates his other roles. In the Mexican cave of Oxtotitlán, David Grove found, painted in splendid Olmec style, a very grand personage seated on a monster's head and wearing a cutaway owl mask probably symbolic of rain. In the same cave there was a smaller painting, an Olmec face also with a cutaway mask, a serpent mask. This, however, covered not the head but only the mouth.

The same sort of curious mouth shield has been found among burials of somewhat later times at Monte Alban. Here, cast in gold, are many small masks of Xipe and other gods. The masks are themselves shielded, lip and cheek, by skeletal jaws and stylized golden swallows cunningly wrought. Are these meant as outsized nose pendants, or do they confer yet another identity?

Southward through Central America and into Peru we find similar objects, masks to be hung from the nose, masks which conceal the mouth. These are cut in metal foil, often gold, and are fringed into stylized cat's whiskers which, on closer view, prove to be writhing snakes. The specialist calls these "kennings"—stronger, more explicit metaphor. In the burials of the mighty we find death masks provided also with mouth masks, two in one. Is a dual statement to be here inferred, or single meaning, subtly qualified?

Keeping cool

Noting the many instances through the Americas of mouth concealment, one remembers of a sudden a very different sort of study, one made many years ago by the anthropologist Robert F. Murphy. He wondered why it was that, alone among Islamic folk, alone among pastoral nomads of the Sahara, the Tuareg veil their men, not their women. The "Blue Men," so called for the indigo dye which rubs off on their skin, never remove the facial cloth attached to their turbans. They may loosen it a bit in the presence of wives and close friends, but its removal is unthinkable. Cigarette smoke filters through the veil. Tea is artfully maneuvered under it. It is a fixture throughout life. Very necessary, they say, to protect the face from blowing sand. But women, as often in the open and equally vain of their skins, do not veil. Why is that? Dr. Murphy wondered.

Then he examined the pattern of daily life among the Tuareg, the rivalry for prestige, the stresses of residence—with the wife's people, for the Tuareg, again unlike most other herding folk, reckon descent through females. He noted the times when men tightened their veils, leaving only a glitter of eye to be seen. Always this happened when a man was in the presence of a social superior or when confronted by his in-laws. And Murphy concluded that the veil was indeed a face-saving device—not from desert winds but from the winds of criticism and disapproval. By tightening his veil, a man kept his cool.

We think of eyes as windows of the soul, and some of us, eager to keep our souls to ourselves, wear "shades"—dark glasses—indoors and out. But eyes reveal nothing. It is our mouths that betray feelings to the world. When we are hard pressed, full of fear, and more fearful still of letting it be known, our fingers go automatically to the lips, curling about them protectively, holding their changes firmly in check.

Watching films made by Napoleon Chagnon of the Yą̧nomamö, those Orinoco Basin people who to this day pride themselves on their ferocity, one is struck by the frequent use of mouth gestures and mouth concealment. For the Yą̧nomamö male who expresses fear or even a slightly diminished confidence is no man at all. And so unless he is himself declaiming, holding forth to friends and enemies, gesturing expansively, he subsides, on haunches or in hammock, and places hand to mouth. He is hidden.

Perhaps ancient men of the New World took the gesture one step further. In a world where rivalry was intense, where rank could be attained by wearing a poker face, the mouth mask must have been a very present help in time of need, a blessing to the man whose expression could give the lie to his words, however much his voice rang with conviction.

The mask of evil

The mask saves "face" and gives "face." It confers rank and an identity larger and more grand than that which is rightfully one's own. It can also provide the anonymity essential to evil, to atrocious acts dissociated from everyday norms. Executioners of all times and places have worn masks and wear them still. So do men bent on using terror as a weapon.

Africanists have long known of secret societies in the Congo region and in the Cameroons dedicated not to order and social control but to sorcery and, more than sorcery, to murder outright. Such men have worn masks of leopards and of lions, have left along their traveled paths the pug prints of felines made by carved simulations thereof. They have killed their victims with "claws" made of iron, knowing that the mute "animal" evidence would sow fear among their fellows and make the leopard-men all the more secure.

Our world knows well such leopard-men who kill and maim and are secure. They wear ski masks and Tuareg-like veils and tight silk stockings which flatten and distort the face. While most of us struggle to adjust our natural features to suit each of our changing job descriptions, our varying roles in life, some of us elect to wear the mask of concealment, licensed by it to commit crimes impossible while exposing our own faces to the world.

To wear that mask is to suspend all other identities—the tender lover, faithful friend, dutiful son. To wear it is to absolve those other selves of guilt and shame. To wear it is to sanctify evil and give it rein.

No animal kills for pleasure, concealing as it kills. *I am the mask; therefore I am human.*

Sadly, yes.

PART TWO

BODY SUITE

5

FLUIDS AND EMANATIONS

Wetness as Metaphor

"I have nothing to offer but blood, toil, tears, and sweat," Winston Churchill told the House of Commons in the terrible early years of World War II. The British people, recognizing a heroic phrase when they heard one, repeated it as "blood, sweat, and tears" and prepared to give their all in defense of the realm.

It was a heroic phrase, and it galvanized resolve precisely because it resonated chords of sentiment beyond the reach of mere eloquence. By invoking the body fluids, Churchill had touched the most elementary levels of existence and stirred the very wellsprings of life. We are, after all, mostly fluid—shake us and we slosh; drop us and we splatter. We are fluid-powered and fluid-impelled. The by-products of everyday physiology are fluid. So are the by-products of emotion. And sometimes it is hard to know where one leaves off and the other begins. In the waters of life there lurk meanings powerful and deep. But the meanings are not everywhere the same.

Churchill chose the right waters to symbolize British courage and summon forth British will: the blood of sacrifice, the sweat of labor, the tears of bereavement. It would not have occurred to him to use others in such a context. Urine, saliva, and semen may be euphonious, but they carry, in our world, a negative charge of meaning and are virtually unmentionable outside the doctor's office. They are, nevertheless, power substances in other cultural settings where the grand Churchillian triad might fall, in terms of symbolic content, altogether flat.

Tears of affliction and blessing

Of all the Churchillian fluids, tears would seem the most universal in meaning. Where there is sorrow, there are tears. Where there is pain, one weeps. But which one? In the Western world (or at least in its Anglo-Saxon component) men are denied the luxury of tears. Only females may indulge in "a good cry." Elsewhere the pattern is reversed. Arab men may weep buckets and in public; women may not. Among pious Hindus of the Brahman caste, the woman in menstrual seclusion is positively forbidden tears, however miserable her condition.

Nor are tears universally associated with sorrow. Andaman Islanders of Radcliffe-Brown's acquaintance wept in welcoming long-absent friends and for every other social occasion, be it merry or sad. As if spigot-controlled, the tears gushed and ceased, prompting the good professor to describe Andamanese tears as a kind of social glue.

Many dwellers in arid lands have seen in tears the body's analogue to rain and have used the one to command the other. Appeasement of Tlaloc, the rain god, required Aztecs yearly to sacrifice babies by the score, and the more they cried, the better. Because of those sacrifices, because of the Aztec meaning of tears and association of babies with tears, we can interpret the art and iconography of much earlier Mesoamerican folk. We can understand why figures of bawling infants turn up in every Olmec site and why every Olmec face has a downturned, pouting mouth. For another ancient and extinct American culture, for the empire of Tiahuanaco, Inca forerunner in the Andean highlands, there is no such hindsight. We do not know why the god of these people was always depicted shedding tears, real tears or tears represented by figures of snakes and birds. This is what we see on painted pottery and carved in stone. Clearly the god's tears have meanings powerful and strong. More than that we cannot say.

Perspiration and what comes after

So does sweat have meanings powerful and strong, though the mention of "strong" in this connection brings to mind a connotation more of distaste than of magic. For we have ambivalent feelings about sweat. The proverbs and sayings of the work ethic glorify sweat: "honest sweat," "by the sweat of his brow," "sweat in a good cause" (Churchill again).

The mini-dramas of commercial television, hawking a vast array of deodorants and antiperspirants, convey a rather different message. "If you must work," they seem to be saying, "at least don't smell like it!" Sweat, honest or otherwise, has come to be both an economic and a romantic liability. And Americans, valuing as they do popularity as much as success, erase it or prevent it with anxious care.

Among other peoples, stopping wetness is a concern only where sorcery is a concern. A well-sweated garment can be used in conjuring its owner. For body fluids, elsewhere, are never simply substances, impersonal and without life. Separated from the body, they still belong to it, carrying everywhere the essence of self until they have vanished quite away. The watchword, therefore, is caution about what happens to sweat, not about sweating itself.

Only one man among the Dogon worries about wetness. Considering that this West African tribe inhabits the high, hot hills overlooking the Niger River, that would seem a vain concern. To the Dogon chief, the Hogan, it is more like a sacred obligation. He is, for his people, their intercessor with the gods; his body their guarantee of being. And so he must neither sweat nor wash, for water rolling on his skin would carry away vital moisture and diminish Dogon life.

Sweat in still other social settings may be something not to be prevented or guarded or neutralized, but to be induced in specific places and at specific times. With very few exceptions, Indians of North America and Mesoamerica were addicted to the sweat bath. Indeed, the separate sweat house—tent, wood hut, hole in the earth, or grand stone edifice—marked nearly every community. Often the sweat house did triple duty as a place of ritual and the local men's club. There the male fraternity sweated before an open fire or around heated stones doused with water to make steam. They sweated for the good of their bodies and the good of their souls. They sweated prior to competing in ritual ball games and before and after difficult undertakings. They sweated for strength, and they sweated for valor. The men of at least one Indian tribe sweated for wealth. The Yurok, a hunting-gathering-fishing people who lived in what is now northern California, believed fervently in clean living and self-denial as the way to achieve success. And for them, success was affluence. Whoever said wealth had to mean money? Dentalium shells served the Yurok just as well, along with fine flints, white deerskins, and the red scalps of woodpeckers. Proper sweating was thought to steel the hunter's arm and sharpen his eye, thus ensuring proper accumulation of wherewithal. Among the Yurok as among ourselves, God helped those who helped themselves.

Today's tribesmen in the eastern New Guinea highlands periodically convert the local men's house into an equivalent of the sauna and hole up in it for a five-day stay. This is easy to do because throughout much of New Guinea men and women live apart, wives in their own houses, men all together in bachelor digs. The avowed object of the sweating ritual is an increase in male power and strength. And all of the attendant austerities—the abstinence from sexual congress and certain foods, the bleeding of noses and tongues—underline its importance. But there are, nevertheless (as Ronald Berndt points out), flirtatious and even competitive overtones.

Men eye one another (covertly and with some jealousy) as

sweat forms designs on the skin. The clearer the design, the greater the warrior: sweat has spoken. Women watch, too. For they are allowed in and out, bearing food. They find the sweat designs exciting, long to have the imprint transferred to their own skins, and sigh with passion—all in vain. Sexual byplay of any sort will remove a man's sweat designs quicker than a bath and spoil whatever chance he has to one-up his fellows. Women may look but must not touch.

Sweat has many meanings, both symbolic and magical. It has many physiological functions as well, ranging from the evaporative cooling of the body core to the release of certain odors, once perhaps delightfully erotic, now to be erased or masked or both. Early in mammalian evolution, various sweat glands were retooled for the production of milk.

The waters of whiteness

Unlike sweat, milk is fluid with but a single function and little variation in symbolic content. It is something which binds mother and child and is, therefore, in societies which count descent through the female, the fluid symbol of the line. Of all things white and pure, mother's milk is surely the purest pure and the whitest white. It appears, moreover, in a context of love and nurture, utterly without nuance. All these givens go far to explain associated color meanings. (See Chapter 1.)

Semen is also white and pure, also nourishing. "Much is needed to make a baby," say some peoples. "Much is needed to make the fetus grow," say others. "Too much," holds yet another group, "makes for multiple births, and who wants that?" Often, in tribal lore, the two white, nourishing substances are thought to be in opposition, and one spoils the other. Such a belief places the nursing mother off limits to Dad and achieves effective birth control.

The nature and function of semen, it must be admitted, is everywhere less well understood than that of milk. (This is true in some segments of our own society, where a female's tender years and the use of Coca-Cola douches are thought to prevent conception.) It must have taken humanity quite some time to tumble to the possibility of biological paternity. Or so specialists have liked to assume. Stories from ancient writings together with notions cherished by remaining tribal innocents have lent support to this view.

No land mass has harbored more innocents than Australia, where conception is believed by Aborigines to occur through action of baby spirits—disembodied clansmen awaiting womb reentry. Sometimes such spirits, en route to a receptive female, are thought to appear first in Father's dream, soliciting directions. Next morning, the proud papa-to-be cheerily announces the happy news: "Guess what? You're pregnant." The Tiwi of Melville and Bathurst islands long ago elaborated this basic belief into the very fabric of social life. There were no spinsters among the Tiwi, no widows, no demure maidens to be courted and won. Every female from cradle to grave had to be married—actually or in name only. For who knew when the baby spirit would strike, and no child should lack a male protector.

This charming "just-so story," it now begins to appear, may rank, even among the Aborigines themselves, with our tales of storks and cabbage patches. Perhaps it has been a story for anthropologists. Professor M. J. Meggitt, living among the Walbiri of Australia's central desert region, found that explanations of conception depended entirely on the time, the place, and the sex of his informant. In ritual situations, men insisted that spirits caused conception. In secular situations, they credited spirits and semen equally. The women, ever practical and excluded from ritual, put their faith in semen and never mind the spirit crew.

Milk and semen did not and do not figure much in Western symbolism or Western conversation. They are virtually banished from the folklore of broadcasting. (Imagine: "Make your semen its whitest-white with doctor-tested Yohimbin!") For a century, mother's milk could not be mentioned because of its anatomical source, a zone alluded to only with the most circumspect euphemisms (except at stag parties, of course). Today, when almost anything goes, it seems hardly worth mentioning and is quite neutral in meaning. Most new mothers prefer feeding their babies a commercial substitute.

Other waters

Another fluid banned from the airwaves is urine. There is an oblique approach, to be sure, through the ads for diuretics. So firmly do these focus on aching backs and overexertion, however, that the uninitiated might never learn the real purpose of the drug.

If television advertisers and executives are reluctant to mention urine, the United States Army is not. Its instructors in field survival techniques recommend the use of fresh urine, sterile at the source, for the cleansing of implements when water is unavailable. The same and other practical uses of urine are well known to tribal folk, who have washed everything from wounds to hair in it and then have tanned leather in the leftovers. Countless are the remedies, ancient and modern, with urine as the star ingredient. Egyptian doctors, so Herodotus tells us, preferred the "waters of a chaste woman." The Ndembu of Central Africa favor the urine of an apprentice circumciser. The Right Honorable Morarji Desai, former prime minister of India, daily drinks his own. He extolled the curative virtues of this regimen on the CBS news program "Sixty Minutes," while a considerably rattled Dan Rather declined to give it a try. (Difficult to ban a prime minister's comments.)

Custom may dictate the place of urination. Never, among most people, in a sacred area unless the water is to be ritually used. Custom may dictate the time of urination. Never when one is concentrating one's thoughts, say the Toradja of the Indonesian Celebes. Much too distracting. Seldom, however, does custom make of urine the fluid of imprecation. The power to curse is more often reserved for saliva. Certainly this has been true in the West.

Salivary signals

"I spit on your mother's grave!"

"I spit on the heads of your children!"

"May you be cursed forever!" When the curser aims his spittle at the offending face, the expression of contempt is ultimate.

If saliva seals the curse, it can remove one as well. The Greek peasant, hearing her grandchild overpraised, mutters softly, "So it should be clean," and turns aside to spit. She thus averts the evil eye. For in excessive praise lurks envy, and envy withers and destroys. Among the Greeks, spitting is an act of power and dread, unclean and therefore prophylactic against filth. Not for frivolous reasons is saliva allowed to escape the lips.

No wonder, then, that Greek immigrants, arriving in America during the early years of this century, were horrified to find tobacco-chewing in general vogue and its by-products not always deposited in the nearest available spittoon. If streets were unpaved with gold, they certainly bore the stains of tobacco juice. "Chewtabaccia," the Greeks disgustedly labeled their hosts. And though chewing gum has effectively replaced tobacco as a favored cud, the original term continues in use when Greek-Americans want to separate the ethnic sheep and goats.

For sheer horror of saliva, no people can outdo traditional Hindus of the higher castes. Nothing that can possibly be contaminated by anyone else's saliva is permitted to touch the person, much less be ingested. Leftover food is fit only for the hopelessly unclean. Dishes and tableware used by others are off limits. Even one's own saliva is dangerous and should not touch the fingers. Thus food is tossed into the mouth, and water is poured from afar rather than drunk. Diners become, so we are told, quite adept at this sort of juggling.

In saliva the Toradja find a sort of all-purpose power fluid of limitless symbolic value. With it one can express contempt but also banish bad omens; one has only to spit in all corners of the house. Saliva figures in medicines internal and external. For sore joints and legs that have "gone to sleep," it is a sovereign rub. Connected as saliva is with the power of speech, the act of spitting lends weight to one's words. In this belief, many peoples have concurred:

> The Toradja, in court: "I am reconciled with you, mine enemy. Do you therefore reconcile yourself to me."
> (spit)

The Maori, among his peers: "I swear that what I say is true."

(spit)

The Creek in council: "I wish to address you, my fellow chiefs."

(spit four times)

Here the act has both symbolic and physiological significance. Going dry in speechmaking, whether from anxiety or stage fright, is a well-known hazard of the platform and the reason for the glass of water by the lectern.

For many of the world's people, spitting is a benediction:

> And [Jesus] took the blind man by the hand, and led him out of the town; and when he had spit on his eyes, and put his hands upon him, he asked him if he saw ought . . . and he was restored, and saw every man clearly.
>
> (Mark 8:23–25)

In the cosmologies of ancient Egypt, wise Thoth is said to have spat into the empty eye socket of Horus, the sun god, to restore his vision.

Healers of the Far East, from the bearded Ainu of northern Japan to the urban practitioner in modern Thailand, bless and cure by spraying patients with water from their mouths. This is true as well in North America, peopled long ago by an Asian contingent. The archbishop of the Ethiopian Church regularly sprayed his congregation. Fathers in the east New Guinea highlands spit in their sons' mouths to ensure growth and, for fertility's sake, spray their pubescent daughters with well-chewed salt.

Indians of the American Southeast were even more explicit. Saliva was, for them, not only benediction but the very water of life itself. It was the bodily echo of that great underworld river, tumultuous and disorderly, from which flowed change and creative purpose, the fertilizing urge of the universe. As the white waters of earth connect and merge, so saliva was thought to join and connect bodies in a social skein. It further joined the inner self to the outer world through its associations with breath and speech. Ideally, one repaired to the river to pray. If this was for some reason impossible, one could achieve an approximate state of grace by spitting into one's hands and applying the moisture to body and face.

Yet saliva, as Charles Hudson of the University of Georgia points out, was opposed to blood, as the water of the lower world opposed the heavenly fire. One did not mingle blood with water, and extinguishing a burning brand in the river signified death. Certainly one did not spit into fire or throw into it any saliva-bearing material. This prohibition appears among many peoples sensitive to the elementary polarities of life: fire and water, red and white, hot and cold.

Missing in the sequence are wet and dry. Blood may be red like fire, hot like fire, but it is not dry. It is another river of life, a vital fluid to be elaborated into metaphor of the sort invoked by Winston Churchill. As a life principle it has deep meaning in our world. That the figure of the vampire carries such horror (and enjoys such popularity) relates to this metaphor and to the theft of life at its source.

The dual meanings of blood

Of all the fluids, however, blood is the most ambiguous in symbolic content. For there are, after all, two kinds of blood. There is the blood which is caused to flow (with wounds and sacrifice). And there is blood which flows without cause, mysteriously and unbidden (menstrual blood and the blood of childbearing). One is "good." The other is "bad."

Toradja in the old days considered some of the "good" kind not really good enough, certainly not worth the spilling. Incestuous criminals and disobedient slaves were regularly dispatched by means that avoided blood. Too contaminating, it was thought to be. Otherwise blood was and still is a strengthening substance, to be applied to everyone and everything: the sick person, the nursing mother, the baby's new teeth, the rice crop, the rice seed, the place of clay, the new house, the earth itself. Chicken and pig blood will do; human blood is better; enemy blood is best of all. In the old days, when the Toradja were still headhunters, long cease-fires were customarily broken by polite invitations to an enemy to fertilize their joint crops. Everybody knew that for a declaration of war.

According to N. Adriani and Albert Kruyt, Dutch anthropologists reporting in 1950, the Toradja warrior cleaned his sword by licking it, an act thoroughly comprehensible to ancient fighting people the world over, people for whom blood is both occupational by-product and potent symbol. Thus the Scythians of Central Asia sealed alliances in blood, solemnized oaths in blood, gave witness in blood, and poured it over their most sacred object, an ancient weapon of iron.

Australian Aborigines spill only their own blood and consider it a sacrament. It is the duty of every adult male to prepare his arms for ritual bloodletting by binding them tightly with ligatures. Arm blood is the universal stickum in making costumes for ritual. Arm blood anoints the sacred objects and shrines. In arm blood are the Dream Time heroes summoned and in arm blood celebrated.

The initiation of young males, equivalent as it is to a rebirth, requires blood from yet another source. Initiates to male cult life are circumcised. Later each new-made man is expected to request subincision, an operation in which the penis is slit right through to the urethra. Whenever baptismal fluid is required for initiates, the subincision scars are reopened and the frightened young men blessed and fortified both inside and out.

The notion that blood is graded in sanctity and healing properties according to the place from which it is drawn turns up in the lore of many people, among them the ancient Maya, who may have practiced subincision for the procurement of A-1 blood. Certainly they were not restricted to one anatomical location. Arms, noses, tongues, cheeks were pierced with stingray spines. Ears came in for a lot of attention, as they did throughout Mesoamerica. Spanish conquistadores described the ears of Aztec priests as slit to pieces, and their clerical garb as caked with blood. Certainly the Mesoamericans were not averse to shedding blood, theirs or anybody else's. Nor were their gods. Having given celestial blood to sustain life on earth, the gods required human blood to ensure their own continued existence. The blood of sacrifice is an old concept, one not unknown (though much attenuated in meaning) in a Christian context. "Take ye and drink," said Jesus, offering wine to his disciples, "for this is my blood."

East New Guinea highlanders studied by Dr. Berndt know a lot about sacrifice. Next to the Australians they are probably the world's most enthusiastic bloodletters, choosing the world's most exotic and (to us) unnerving sources. Tongues are scraped, nasal membranes are rasped, nasal septums and fingernails are pierced, penis foreskins are slit, and the urethral tube linings repeatedly bruised with rods of spiny leaves inserted and twirled. They call this a necessary libation to the ancestral spirits. They call it a powerful means of ensuring strength. They also call it "male menstruation."

Mind over matter

"Anything you can do we can do better," the New Guinea men seem to be saying to their womenfolk. But also, "Things happen to you. We *cause* things to happen to us. No animal can do that. No woman can do that."

"Oh, yes, we can," New Guinea women long ago retorted and promptly began to imitate the men who were imitating them. In their rituals of puberty and every month thereafter, women bleed their vaginas as men bleed their noses. For strength, they say. For defiance as well, it would seem. Typically, a woman takes to her bleeding stick when she is angry, when she wants to threaten and posture and intimidate.

Thus the east New Guinea highlander, man or woman, plays out in body fluids the dramatic antithesis of nature and culture. In his assaults on the flesh, a man defines what it means to be human. And being human means to be in control, to be something different from nature and something more—"a little lower than the angels," as the Old Testament Psalmist put it. In the West the difference was an article of faith until 1856, until Darwin, when humankind began to sink from something more than animal to nothing but. The arguments generated then, the controversies about human attributes and their place in the scale of life, continue still.

That men in many cultures create ritual analogues to menstruation and parturition, hedged around with protocol and rules, suggests heroic effort, not so much to defeat horror of the unknown as to constrain what is natural and potentially powerful, to impose on it order and will. Perhaps this explains why, in the myths of so many small-scale societies (and some not so small), the civilized "firsts"—fire, strength, magic, song—are all stolen from females. The east New Guinea highlanders say they stole special flutes from an ancestress and that, ever since, women may not see these flutes on pain of death. If one gives credence to phallic symbolism, one might see this act not so much as a theft of song as a theft of the instrument of manhood.

An ever-present danger

Whatever the original rationale, women in their moons or in childbed (or sometimes just by virtue of their sex) have been, with very few exceptions, objects of fear and contamination. They have been considered dangerous to the world or to men or to themselves. Sometimes a combination of all three. They have been thought capable of weakening men, infecting their own children, offending the gods. They have been thought to put in jeopardy the hunt, the crops, war, games, journeys, and gambling. And so they have been everywhere required to practice avoidance. Coition is a baseline taboo. To this can be added other proscriptions: certain foods, the implements of subsistence (fishing tackle, hunting gear, fields, barnyard animals), and the things of ritual. Throughout Southeast Asia to this day the menstruating woman is forbidden entry to shrines, temples, sacred springs. The notices are usually printed in English, a good indication of the nationalities from which defilement can be expected.

Often the list of taboos becomes so extensive as to remove the woman from everyday life. She must take refuge in an isolation hut. En route, she must avoid doorsills and balconies lest she contaminate same. Once in the hut, she dons special clothing, eats from special dishes, and refrains from washing or scratching herself. She might otherwise blast her own body. (North American Indians thoughtfully provided special scratching wands.)

Pious Hindus of other days required the menstruant to avoid tears, thoughts of religion, and thoughts of her husband. So threatened was this gentleman that he avoided travel and serious undertakings during his wife's seclusion. Poor folk, lacking isolation huts, turned their wives into the street for the duration of their defilement. Among Hindus that was four days. The number varied up to ten, depending, one thinks, less on physiology than on the numbers considered magically auspicious. It is just possible that the Moslem institution of purdah (literally, "the curtain") is an extension to a woman's whole life of an original menstrual quarantine.

Of the few people who decline to extrude the menstruant, only the Toradja made a special virtue of her condition. Women were thought to be exceedingly brave and valiant, never a question about that. Men, on the other hand, appeared to feel some concerns about the state of their fortitude in battle. The ardent swain, therefore, begged from his lady love a token. No dainty glove, no chiffon scarf would do. It must be the hem of her barkcloth menstrual skirt, stiff and stained. If she really loved him, she might allow him to go right to the source, collecting a small stain on his finger. This would then be carefully protected, carefully bandaged. If ever in battle he felt a twinge of fear, he had only to snatch off the bandage, sniff his token, and pitch into the fray with furious courage.

How Toradja women in their moons managed to acquire such an image, one cannot imagine. In various other settings, however, despised menstruants have learned to manipulate the menfolk who were manipulating them. The Chipewyan woman, living in the snows of far northwest Canada, managed to confuse her husband about the real nature of the menstrual cycle. Whenever his attentions were unwelcome, she simply moved to the seclusion hut. And if that happened several times a month, he was none the wiser.

In the east New Guinea highlands, it is not so much the blood of a menstruating woman as it is the condition itself that renders a woman dangerous. Food given by a menstruating woman or an adulteress can cause her husband to sicken and die. He depends on her, therefore, to warn him about contamination, and if she loves him, she does. Otherwise, no. The message, says Professor Berndt, is clear: "Treat me right, or you'll be sorry!"

The man who suspects contamination of any sort gets out his trusty vomiting stick—a sort of flexible cane, the two ends of which he holds in his hands—and heads for the nearest stream. Since his first pubescent "swallowing of the stick" (an act which some groups swear breaks a hymenlike tissue and ensures sexual unfolding), he has been following this regimen designed to protect and enhance his physical perfection.

Mind over digestion

Now most of us find the act of vomiting so unpleasant that we cannot imagine, cannot want to imagine, symbolic significance in the rising gorge. But a good many people have vomited for health and grace and doubtless will continue to do so. The Papago of Arizona swore by it as a rainmaking device—never mind your sweat or tears. The Southeastern Indians found it an extension of saliva and vomited enthusiastically (preferably in running water) before and after every major undertaking. The council rooms of their ancestors, excavated by archaeologists, reveal basins situated before every chiefly seat. Whether the object in vomiting was to achieve purity, to demonstrate purity, or to foretell the future, says Hudson, is not entirely clear. (Stomach contents, deposited in river water, were thought to be divinatory.)

Certainly the Southeastern Indians were very concerned about food: that it not be bewitched or otherwise spoiled; that it be ritually permissible and thoroughly cooked. They shuddered at raw vegetables, considering them food fit only for animals or the human so lazy as to eat like them. Vomiting provided assurance that what had gone into the stomach remained undefiled, a condition less susceptible of proof when food exited by the other route. One senses here a wish to ignore the excretory process.

East New Guinea highlanders would like to bypass it outright—one of those beastly natural functions, don't you know, uncontrolled and uncontrollable. The men of myth, they say, did not defecate at all, relying instead on their vomiting sticks.

Whatever the mythological wishful thinking, the products of excretion are daily realities. Like the fluids, they are accounted by many folk to retain the human connection. The Toradja say that the child who eats his own feces also eats his life. Feces are considered subject to sorcery and sometimes also fertilizing in ways that have nothing whatever to do with crops. African kings worried about this while on their rare journeys to other tribal lands. Since they were, to a man, accounted sacred, their body products were sacred also. They took pains, therefore, to send these, leaf-wrapped and secure, back to the homeland. Why, after all, should the land of foreigners be so enriched?

Cleaning house inside . . .

For most people, disposal rather than use is the problem uppermost in mind. And the speedier the exit from the body, the better. Purges and enemas are common to every medical lore. The Egyptians, said Pliny, learned the art from the ibis, a bird thought to use his own bill as a syringe. It was the belief then that enemas removed "poisons" from the system and should be employed on a regular basis for health's sake. Merely substitute "acids" for "poisons," and one discovers the same beliefs flourishing today among purveyors of the "high colonic" and other medical cultists. Only flush out the "bad," they say, and glowing good health, a jolly disposition, and all other desiderata will follow. Interior garbage must not be allowed to collect.

Even among the general public it is a familiar concern, one which the most casual television viewer could hardly ignore. If variety in brand names is an indication of value, then laxatives should certainly tie automobiles and beer for first place in public interest. Sensitive as we are to the mention of other body fluids, we can, when it comes to purges, tolerate quite a good deal of explicit huckstering. We hear of the laxative that is a "true, blue friend." Another is "good medicine at a good price." Another restores "normal regularity." And always at dinnertime. The very extent of the coverage should alert us to meanings that go quite beyond those of functional efficiency.

Fifty years ago, before the psychologist warned of rigid toilet training and the pediatrician forbade purging, children were frequently dosed with syrups of this and syrups of that and told that the ensuing cramps were good for them. Some nostrums were reserved for special occasions. Journeys, for example. Before driving to the hill country or the seashore, the careful family underwent a "course of cathartics." This involved three days of misery and a departure in dishrag condition. By today's wisdom, none of this makes very much sense. As ritual, the course of cathartics makes all the sense in the world. Travelers, whether they knew it or not, were warding off dangers of all kinds—foreign influences as well as foreign germs.

Another familiar ritual of those years was the Saturday night bath and accompanying dose of castor oil. One needs to remember that those were days of fewer bathrooms and more people per house. Also, the demise of God had not yet been proclaimed. One prepared for Sunday by thorough cleansing, inside and out. Cleanliness, after all, was next to godliness.

That has been, and still is, a tenet of every faith, though one often needs to examine just what is meant by "clean" and which kind of "clean" counts. Outside or inside? Freedom from dirt? From germs? From sin? From deceit? ("You'd better come clean!" the district attorney threatens. "I'm clean, I'm clean!" protests the suspected thief.)

. . . *And out*

And what makes one clean? In every faith: confession, ablution, anointment, catharsis. Substances used in the last three acts vary widely. Among Hindus and cattle herders of East Africa—both places where the cow is a sacred being—fluids and body products of the cow have uses both practical and ecclesiastical. They figure in firemaking, face-washing, anointment, and purifying beverages. The Masai pastoralist plasters the cut umbilicus of his newborn babe with fresh dung. The Brahman uses it to purify his place of rest.

Water is everywhere the cleaning agent par excellence, but

its powers may often lie more in symbol than actuality. The holiest rivers in the Hindu world, the most sacred baths, the spots most surely proof against contamination, may be, in the medical sense, alive with filth. The pious Moslem is adjured to wash five times a day and may use sand as a substitute. The Christian is baptized with water, and sometimes in it if he belongs to a sect requiring total immersion, and nobody frets about pollution. Fonts at the door of every Catholic church may be holy but surely could not win honors for hygienic properties. Obviously, freedom from dirt—earth, germs, dust—is not all that matters in being clean. Freedom from sin, from defilement, from contamination count for more.

This may be true even in circumstances with no obvious religious point. It may be true of baths taken with hygiene as the express purpose. It may be true when the cleansing agent is 99 44/100ths percent pure and the water certifiably safe for drinking. The bath must surely have hidden meanings for a population that indulges so frequently as to make "dry-skin itch" a hazard of the wintertime. There is the story of the man who complained that his wife regularly showered after work, at bedtime, and again before work in the morning. How could she get so dirty, he marveled, working in an air-conditioned boutique, living in an air-conditioned apartment, and sleeping in a bed made up daily with fresh sheets? How, indeed? Why did the Egyptian priest bathe twice each day and twice each night?

In water there is grace, and in flowing water still more. What the river was to the Cherokee, the east New Guinea highlander, the Egyptian, the Hindu, the bathroom spigot is to us. It is under the spigot, in our bathrooms, that we "come clean"—whatever that means to each of us individually. It is in the bathroom that we deal with body fluids—with the excretory products, with blood ("good" or "bad"), with sweat, even now and again with tears. Fluids are observed there, examined, treated, and removed.

So important is the bathroom in American culture that it can determine the worth of a house or apartment. Owning multiple bathrooms is as much a status symbol as owning a high-priced car. Kitchens we can sometimes do without (especially with the multiplication of fast-food outlets). The well-equipped bathroom, never. Gleaming, antiseptic, odor-free, everywhere of identical design, the bathroom looms large in architectural convention, as it does in the American value system.

No wonder Professor Horace Miner, in his spoofs of anthropological investigation, chose to focus first on the bathroom. How, he wondered, would an ethnographer fare among those unknown, primitive, heretofore unstudied people called the Nacirema? (Try spelling it backward.) What would he make of that special room in all their houses, of its cabinet full of magical potions, of the private and secret rites performed there, of the anxious initiation young children received there? Obviously, said Miner, he'd make of it a shrine.

Why not?

6

PARTS AND PIECES

The Anatomy of Meaning

"Let's get the body out of weights and measures," say the masters of science and the engineers. "To reckon in feet and hands, in inches based on the breadth of somebody or other's thumb, in miles based on the thousand strides of some forgotten Roman soldier—all this is archaic, inexact, and unstandardized."

Learned Americans have been mulling the problem since the 1790s, when the metric system was introduced in France. They are mulling it still. For though ordinary Americans may have relinquished over the centuries their ells and furlongs, they still cling to their inches and feet. There is, after all, something comfortable and comforting in those familiar units of measure, some lingering continuity with older times and older ways.

The recollection that the original yard was standardized as the length between the royal nose and index finger of the first

English Henry endows that larger measure with personal content. The same is true of the Egyptian cubit, which spanned the length of a pharaoh's arm from elbow to middle finger. It is a measure which lives today only in the biblical blueprints for Noah's ark. The cubit was a sovereign standard in the engineering projects of 3000 B.C. and is the first known to us in writing. But it was not the first, as scholars are now beginning to discover when they study architectural remains of the Neolithic. In Western Europe and the British Isles, for example, there abound stone circles, tombs, and alignments, many of them a thousand years more ancient than the Egyptian pyramids. The unit of measure employed in their construction has been called a "megalithic yard" by Professor Alexander Thom, late of Oxford University. Not quite thirty-six inches in length, the megalithic yard may itself be the multiple of a smaller unit, something close to an imperial Roman foot, not quite twelve inches in length. People and their extremities were smaller then.

Body yardsticks

Hands, feet, fingers, arms, stride—these were our first rulers, our first surveying tools. How else should we have cut the world down to size than by way of what we already had and already knew? So many fingers laid together equaled a foot. So many men's feet placed toe to heel made yet another unit. What could be walked before the sun stood high gave meaning to distance. People as yet untouched by the measurements of the West still construct their shelters and temples to accord with the plan of the human body and its parts. Thomas Rhys Williams reports that the Dusun of North Borneo have twenty-four such measures of length (fingers, knuckles, arms in various combinations); eleven of height (as short as an ankle to "over my head"); two of circumference (the circle of fingers, the arms rounded as in an embrace).

Even the precise and elegant metric, defined as it is in terms of the earth's circumference, may yet have had its own body beginnings. It is, after all, decimal in nature, each successively larger unit built up by tens. Counting by tens has been around and has been popular for a very long time in human history. One suspects, however, that it would never have been devised by human beings with twelve fingers and twelve toes.

By extrapolating from the body, we have achieved orientation as well as measurement. While not precisely rectangular, we do have six usable dimensions: front and back, the two sides, and the poles of head and feet. They are and have been applied in every language to the description of position:

"Move sideways."
"Best foot forward."
"Go to the foot of the stairs."
"Take a left."
"Where's the head?" (In Navy parlance, the latrine was originally located in the head of the ship.)

Position allows for even more abstract connotations:

"My kid is at the foot of his class."
"Buck up and face the world."
"He's my sidekick."
"Meet the head of the company."

All these have little to do with walking, standing, or turning.

The body as location referent appears just as frequently elsewhere. Among many peoples, such as the Dogon, the plan of a compound and the plan of a village are said to conform to human shape. Joseph Greenberg, the linguistic anthropologist, finds in African languages generally a widespread use of body parts to express position. An item is "the belly of a box" when it is inside. It is "the head of the table" when it is on top. A snoozing wayfarer does so at the "buttocks of a tree," not at its foot.

Locating emotion

While the external human body has left its indelible imprint on the world outside—on measurement, direction, position, orientation—its inner organs, the workings beneath the skin, have given location and dimension to feeling. Not so very long ago we were interpreting states of health and states of mind in terms of the four humors, orginally the sovereign juices of the body. An excess of one or the other was believed to make one sanguine, choleric, phlegmatic, or melancholy. The moods remain, though the juices themselves are no longer considered accurate barometers of temperament.

Human beings everywhere have endowed the various organs with rather different powers and emotional content. Our culture makes much of heart as the seat of devotion ("My heart is full of love"). Grief "breaks our hearts." And, sure enough, there is inevitably a pain in the chest to prove it. The heart encompasses courage of a particularly gallant and noble sort ("You gotta have heart!") and warm compassion as well ("He's all heart!"). It provides ultimate focus ("the heart of the matter"), ultimate concentration ("with all my heart"), and ultimate substance ("a really hearty brew").

But the heart, however strong its beat, is not strong enough in some West African cultures to resist supernatural invasions—sometimes good, mostly evil. Witches are said to accumulate a telltale substance around the heart, where it can be seen in autopsy. According to the Tiv of Nigeria, this substance, called *tsav*, comes in three colors—red, black, white. Your guess as to which is the evil one. The Hopi of Arizona make a similar connection. Their witches are said to possess two hearts, the spares acquired in the disposal of relatives.

For the Dugum Dani of West Irian, the heart is the seat of the life-force, the personality, the individual essence. They call it "the seed of singing." The ancient Egyptian heart was actually equivalent to mind, and the bowels were also involved in the cognitive process. The Yoruba of Nigeria locate in the navel region the ultimate expression of character. Where we might say "He is a good-hearted man," the Yoruba insist that the same man has "a good belly."

We relegate to the abdominal region the more elemental, the more instinctual aspects of our natures. We "can't stomach" something offensive, and rage "roils our guts." We speak of "gut wisdom" when we refer to the primitive substratum of intelligence, and we feel foreboding "in our guts." Courage of the fairly uncouth and reckless sort is to us "gutsy."

Though part of the general abdominal suite, the liver has for us little metaphorical utility. When we describe the coward as "lily-livered," we have gone about as far as we can go with that organ. Among Africans, however, the liver carries many rich allusions. It is considered the seat of moral judgment. "My liver is white" says the Ndembu when he means "My conscience is clear." According to the Sudanic Azande, the liver attracts forces of evil. Azande autopsies focus on this organ, not the heart, for clues to supernatural misbehavior on the part of the deceased. Many people, including the Romans, have considered the liver an organ of divination but have declined to use human samples in this activity. The Dogon make the liver responsible for the highest human aspirations, locating there the seat of the soul.

One would be hard put to find in the Anglo-American body a specific address for the soul. When we sing, "Heart and soul, I fell in love with you," we tend to use heart with a connotation ever so delicately sexual and leave the ethereal ingredients of affection to soul, wherever that entity may be located. Eyes are described as "windows of the soul." Is the soul in the head, then? Not necessarily so. We like to keep it homeless and free-floating.

The sensory suite

The West in general has made much of eyes, vesting in this component of the sensory apparatus a major burden of metaphor. Hurricanes have eyes; so do potatoes, needles, and the loops for hooks. We associate stars and planets with eyes, as in this poem by Francis William Bourdillon:

> The night has a thousand eyes
> And the day but one,
> Yet the light of the bright world dies
> With the dying sun.
>
> The mind has a thousand eyes
> And the heart but one,
> Yet the light of a whole life dies
> When love is done.

In ancient Europe and the Middle East, the eye has been a symbol of divinity since Neolithic times. Inscribed on ancient potsherds and carved into the stones of megalithic tombs are the swirling circles and sun-ray lashes of eyes, watchful and disembodied. Sometimes the eyes are localized in schematic female faces. At the Mesopotamian temple at Tell Brak, dating to about 3000 B.C., excavators found countless little female effigies, all bodies surmounted by eyes and not a face in a carload. Surely a graphic reminder to hard-pressed folk that Mother saw all, knew all, and would provide.

For all its association with divinity, the eye in lore of the Middle East and of Middle Europe also carries evil connotations. It has been and still is thought to be the high-powered transformer of witchly designs and, as such, is "the evil eye." The Greek peasant holds that green-eyed people are likely to be witches whether they know it or not. Azande tribesmen suspect the person with eyes red and inflamed. The Hopi in his pueblo calls the medicine man gone bad "Eye-walking-doing."

Among the Suya, those central Brazilian folk earlier described, there is nothing ambivalent about the eye. As metaphor, it is all bad. To say that someone has really keen sight is neither a recognition of fact nor a flattering comment. It means that such a person is believed to possess the strong, wild vision of animals and can thus see everything—the spirits in the sky, the below-earth people, the action in neighboring villages, everything. This evil power enters through the eye of one who cannot "hear."

When a Suya says, "I hear," he means (as Anthony Seeger reports), "I have received the information, I understand it, I have incorporated it into my fund of knowledge." Pretty much what we mean when we say, "I see!" "I hear" usually indicates for us only sensory reception and includes no intention to "see." ("I hear you talking, but you can't come in" used to be the phrase in full.)

The Suya who does not "hear" is considered willfully immoral. And this serious defect of character will render a man incapable of public eloquence, which is the hallmark of Suya propriety. (Since women do not speak in public, the failure to "hear" is of correspondingly less consequence.)

The power of speech is, for the Suya, located in the mouth, as indicated by the enormous labrets which men wear in their lower lips. In the Western world, the tongue provides the operative metaphor. The words that have to do with speech—*language, lingual, glottal*—all are "tongue"-derived. Egyptian priests of the god Ptah used the same metaphor, as James Breasted translates:

> It is the tongue which
> Repeats the thoughts of the heart.

Being flexible and mobile, the tongue is also widely employed in the language of gesture. The protruding tongue serves us in lieu of the cutting remark. Drooping from the Maori mouth, the tongue signifies battle rage and, in older times, was meant to terrify the enemy. Accompanied by a hissing intake of air, the exposed Tibetan tongue signifies extreme deference to a superior, the desire to protect him from contamination by one's common breath.

Head first

Eyes, ears, mouth, tongue—all metaphors of wisdom, speech, power, understanding; all associated with the head. No wonder the head, among many peoples, has been considered special, even sacred. Polynesians generally have counted the head, especially of a person of rank, so taboo that such a one has had to eat his own head lice. The Hopi call the head the reception center for cosmic life, the very tuning fork of the universe. Africans generally center in the head both individuality and volition. "He himself" translates in many African languages to something like "he with his head." We consider the intelligent man to "have a head on his shoulders" and speak of "getting our heads straight" when we intend to reorganize our lives.

No wonder, then, that so many peoples have been moved to honor, to endow with special powers, not only the living head but its corporeal remains. No wonder the heads of loved ones have been the bodily parts most often retained aboveground among living descendants. No wonder people of so many different cultures have found enemy heads worth taking and collecting. "Two heads are better than one," we like to say when problems need solving. It is a conviction shared by headhunters, who have considered the severed article as potent a source of spiritual assistance as the attached variety.

Heads of relatives were preferred in Jericho, a Neolithic city already walled, already ancient, before ever Joshua blew his trumpet there. Residents, during the period before the use of pottery had quite caught on, modeled plaster portraits over the family skulls, fitting out the empty eye sockets with cowrie shells. Half a world away and several millennia later in time, the Maya practiced a similar method of portraiture, clothing the facial bones of their great men in gum. In some parts of New Guinea and adjacent islands, such portraits are produced to this day with cobwebs and clay.

Keeping the head

Heads of enemies have often received similar treatment. If not portraiture, then at least a ceremonial disposition. We can see something of this in a Neanderthal cave not far from Rome. There some sixty thousand years ago an enemy, killed by the cliché blow with a blunt instrument, was separated from his head. Deprived of brains, the skull was then set carefully in a special niche (perhaps initially on a stick) and surrounded by a ring of stones. Trophy it may have been, but surely of a somewhat different sort from the deer heads which crown the hearths of modern hunters. Surely some terrible power was thought to reside in the Monte Circeo skull, a power that might be harnessed for use by the owners of the cave. The attempt appears to have misfired. A landslide soon after sealed the cave until its rediscovery in this century.

We see evidence of ancient head-hunting in archaeological sites of the New World. We hear of it as well. An early report comes from a Spanish conquistador who witnessed the cult among tribes of coastal Ecuador and somehow lived to tell the tale. The techniques of preservation he described were, apparently, similar to those known among peoples of the Peruvian Montaña and Amazon Basin. Of these, only the Jivaro retain the additional skill of head-shrinking, which is actually a mummification of flesh and scalp detached from the underlying skull.

According to the Jivaro, the making of a *tsantsa,* a shrunken head, accomplished three things: it humiliated the victim and his group; it demonstrated the warrior's courage; it added the victim's courage to his own. One must wonder, however, whether the rewards were really worth the effort and terror involved. From the moment of decapitation the warrior believed himself stalked by his victim's soul. His safety lay in undertaking a whole program of abstinence. For sometimes as long as a year certain foods were forbidden him. He had to live apart from his wife and deny himself all comfort. Not until he could dance with his *tsantsa* at a great feast was all danger canceled. All supernatural danger, that would be. Nothing could avert the vengeance raid in which his own head might be forfeit.

Using the head

In ancient Europe, heads were simply prizes of battle and never mind the protocol. The wrangling Celts, who gave no peace to the Romans or to one another, collected heads with enthusiasm, nailing them to the walls of their houses. Distant relatives in Central Asia, such as the Scythians, also reckoned their triumphs in heads. Display was not what the Scythians had in mind, however. They fashioned enemy skulls, gilded or mounted in hide, into drinking cups.

Others among our European ancestors used the severed head in games, one of which became that grand American institution of football. Tracing the sport back to its beginnings, anthropologist William Arens finds the heads were originally supplied by Vikings trounced while on an English raid. Even after heads had been replaced by inflated bladders, the game itself was still called "Dane's head" and still involved a lot of kicking. By the 1800s it had become institutionalized at Rugby School—hence, the name rugby. With American revisions, football. Would it give the practiced punter pause to remember that the ball he is about to kick was once a head? Probably not.

Island Southeast Asians had still other uses for the severed head. Some insisted that their men could not marry without one. Considering the sort of population decline that would attend their wholesale obedience to such a rule, we must suppose it applied only to young men of high social position. Heads were, nevertheless, such status symbols that many a prospective bride is said to have held out for tangible proof of ambition on the part of her intended. And many a suitor was desperate enough to take easy heads, even those of aged female relatives lured into the bush. An examination of the Dusun trophy house in Sensuron turned up, according to Thomas Rhys Williams's estimate in the 1960s, many more bones of adolescent boys and aged females than of mature men.

In some head-taking cultures, the need reaches beyond courtship and prestige to encompass the continuity of life itself. Ancient Chinese records suggest that, during the Shang and early Chou dynasties (ending around 800 B.C.), men of rank acquired names and souls for their sons by killing enemies and burying their heads beneath city gates. Not just any wayfarer would do. The head taken for this purpose had to belong to someone with a reputation, with a name that had preceded him and would do honor to the boy who would henceforth carry it.

Fertile heads

That same custom could be found as recently as 1956 among the Asmat of swampy coastal New Guinea (West Irian) when the Reverend Gerald Zegwaard was the first white man to live among them. He was lucky to avoid decapitation during his sleep, as the Asmat were not above this expedient, so eager were they to acquire named heads. For only with a head and a name could a young boy complete his journey to manhood. Only with an enemy head in hand could his father escape the taunt of womenfolk: "You piece of meat without a soul!"

The severed head taken in a raid was placed between the initiate's legs, pressed against his genitals. By holding it there, by staring at it over the course of days, the boy could be assured of good health and fertility. It was his ticket to sexual maturation. Now, the Asmat depend on the sago palm and its pith. This is their staple food. A man's head, they say, is the fruit of his body, as the skull-like coconut is the fruit of the tree. Where the fruit falls, there grows a new palm. Where a head falls, new life follows. So closely twined were the two events in Asmat thought that severed heads were often hung near trees as an additional encouragement to growth.

Maya priests apparently made the same connection, making of the severed head a symbol of fertility. In the *Popul Vuh,* a book transcribed in Roman letters after the conquest, the decapitation of a great hero is described. His head was placed in a barren tree, which soon thereafter was covered with fruit. "For," held the priest, "the round objects which cover the branches of the tree are nothing but skulls."

Command center of the body though it is, the head balances on a relatively slender stalk. It is an extremity easily removed with the appropriate tools and the victim carefully trussed. Decapitation, not surprisingly, has been a popular method of execution. Other extremities have also invited removal as trophies or in punishment. The severed hand in Saudi Arabia has long been the mark of the apprehended thief and is expected to reappear as such in the new Islamic Republic of Iran. For if the brain directs, the hand effects. It is the *doing* organ of the body.

Hand power

One would expect to find in hands a universal metaphor of creativity, and one would not be disappointed. In the Sistine Chapel is the sublime embodiment of this image. There Michelangelo painted the creation of Adam, painted a majestic deity stretching forth his hand to the recumbent and wondering man. The laying on of hands, the curative touch, is everywhere an article of faith. In the Kalahari Desert of South Africa, the Bushman hunter in trance flutters healing hands over the bodies of his comrades. The Navajo of the American Southwest flutters diagnostic hands to discover what part of the patient is ill and what cure is in order. Jesus cured with a healing hand. And English royalty yearly practiced the "king's touch" as a cure for scrofula, a custom lapsed in recent years along with belief.

Hands speak. Many of us cannot produce an utterance without gestural accompaniment. Sign communication has been developed by many hunting peoples to make possible the silent ambush of game. Gestural languages have functioned as lingua franca in many areas of polyglot tribes. (The North American Plains was one such.) The international sign language of the deaf has brought communication to those who would otherwise be deprived. This particular language of gesture has been successfully taught to chimpanzees and gorillas who are incapable of vocal speech because their brains lack specific sites for this gift.

Speech centers are unique to the human brain and uniquely asymmetrical in location. The development of speech has given one of our cortical hemispheres dominance. And hemispheric dominance is reflected in the hands, one of which is normally dominant over the other. On this point of difference what a vast burden of meaning has turned.

Hands good and bad

Most of us tend to favor the right hand. The left is correspondingly weak and so has come to be associated with what is feminine, deceitful, profane, unclean. In many cultures the disgusting necessities of life are relegated to the left hand. The left hand assists the right but does not initiate action. The left hand may not receive a gift or salute a superior. Women must not use the left hand in cooking, and one eats with the right hand only.

Something of this stricture still obtains in American table manners, which allow one to hold one's fork in the left hand while cutting meat but demand a quick shift when meat is conveyed to mouth. We extend the right hand to shake and, at table, seat the guest of honor at the right hand of the host. It is a position sanctioned by Heaven, where at the final trumpet Jesus will be seated at the Father's right hand, itself raised in benediction.

Right comes to us from the same Indo-European root which produced *rex, rajah, regal, ruler* and which at the outset described a "true, strong guide." *Left* derives from roots that imply weakness and are probably cognate to *leper.* In Latin, *sinistra* was the word for the left hand, never to be used in divination. From this comes our word *sinister,* which connotes malevolent intent and bastard birth as well.

Robert Hertz, who wrote some years ago about the implications of handedness, saw in the favoring of the right hand an expression of the human will to elevate the sacred at the expense of the profane. Whatever the intent, it has certainly involved a sort of cultural mutilation. Not so long ago, American children with a left-handed preference were routinely made to switch. Elsewhere the offending member might be forcibly bound. In human society generally, differences breed distrust. And over the long years, left-handed persons had come to be suspect, had come to be feared.

Just when the fear began, we do not know, but the earliest evidence of handedness comes to us from Neanderthal caves. Human teeth found there show the scratch marks of knives wielded by right-handers in cutting off meat close to the mouth. (Eskimo hunters eat in this manner today.) By Upper Paleolithic times (as we have already seen), cave artists were using the hand as a major symbol, though of just what we cannot say. As a signature, perhaps, as a gesture of reverence, or as evidence of participation—all these explanations have been suggested. Most of the silhouettes are of left hands. Perhaps, say the specialists, artists held the trough of pigment in the stronger right hand. There may be yet another reason.

The painted hand appears in rock art of North American Indians and of Australian Aborigines. Unfortunately we find few living analogies to give us a glimpse of meaning. One of these comes from the Dugum Dani of the West New Guinea mountains. For them, the hand in silhouette and positive image is the most common theme in their sacred red paintings. Such prints are made during the initiation of young boys. Without the direct observation of Karl Heider we might never have known this, for the prints are all those of older men.

In North American art, the single hand, whether painted or carved or cut from sheets of mica, is the emblem of the shaman, the particularly powerful shaman, one likely to turn sorcerer. And often the shaman favors his left hand. It is part of the difference that propels him into the profession. Occasionally we can see, scratched in rock faces, figures of shamans with left hands huge and fingers stretched wide.

We find them in the Old World as well. In the rock art of the

Swedish Bronze Age, perhaps from 1500 B.C., there is a particularly striking scene. Before a gigantic snake stands a man with arms upraised in reverence. His left hand is enormous. And there is something more. The figure is ithyphallic, the penis ostentatiously erect.

Phallic symbols and symbolic phallus

Here is another body part fraught with meanings, some of them lost to us today. The cultural climate of the West (first puritanical, now sexually freewheeling) has inclined us to interpret ithyphallic figures as fertility symbols or as ancient works of pornography. And often such interpretations are the only ones possible. Certainly the literature of the distant past contains ample reference to ritual orgies, to mysteries culminating in "sacred marriage." Such events were visual exhortations to "be fruitful and multiply," and all living things—crops, animals, human beings—were intended to take the hint. Herodotus wrote at length of the phallic festivals of his world. They continue today in India, where devotees of Siva carry genital effigies, large and movable. Classical Greece and later Rome honored the god Dionysus with phallic processions which included action and dialogue later to develop into true comedy. The costume of the leading buffoon was always equipped with an artificial phallus. The selfsame buffoon with the selfsame equipment is alive and well and living among tribesmen in east New Guinea, where bawdy improvisation is all the rage.

The rock art of prehistory is not without its explicit scenes of coitus, meant to give nature a nudge in the right direction. And as Americanist Dean R. Snow points out, sexual prowess was ever an expectable attribute of living shamans, male or female. Archaic American rock figures demonstrate that those role requirements had ample precedent. Unfortunately, a schematic vulva does not explicitly identify the sexual appetites and prowess of the shamaness. One cannot miss the shaman with sexual parts at the ready. Indeed, we see them often disembodied and with spreading wings. Occasionally we see birds with human phalluses, possibly a graphic reference to animal transformations said to be the crowning feat of the shaman's art.

Thousands of years earlier in the French cave of Lascaux, hunters told stories about such a personage. Deep in a "well," very difficult of access, they painted a scene from his life. It includes a retreating woolly rhinoceros, naturalistically rendered, a spear-wounded bison, and, between them, the man himself. We can see him today, a crudely drawn stick figure, bird-headed or wearing a bird mask, and he is lying supine. Is he dead, killed by one of the beasts? Hardly. Rigor mortis does not stiffen all parts of the anatomy, and the man is conspicuously ithyphallic. Can such a condition be maintained in trance? Almost certainly, a physician would say, provided the person in question is adept at achieving such states. Trance states are basic to the shaman's professional reputation, and the man is unquestionably a shaman. The bird-topped staff nearby can only be a badge of office. But why has he assumed this particular posture? Is it a demonstration of the shaman's sexual prowess, and if so, for whom? The animals?

Phallic reverence

That brings us back to the question of fertility symbols. Seldom in the rock art of North America, and almost never in the caves of the European Paleolithic, are ithyphallic figures drawn in clearly sexual contexts. Provocative females are few and far between and, while clearly phallic men are sometimes grouped together, they do not interact homosexually. The stimulus at work appears to be a religious one. Attention is focused on power beings essentially animal in form. Remember the shaman of the Swedish rock, offering reverence to a snake. In the Paleolithic caves we also see honor paid to animals. We see men (and sometimes women) with arms upraised, knees bent in homage. We see men costumed as animals, wearing the masks and skins of bison, deer, bear, birds, perhaps revealing the shaman in transformation. Here are men, the pictures seem to say, in the grip of a sublime experience. We do not know its exact nature, but reproductive fruitfulness is surely one of the lesser themes. (Had the role of semen in conception been then discovered?) If the severed head can be seen as a fertility metaphor, why not the erect phallus as a symbol of religious ecstasy?

Alexander Marshak, one of the keenest students of Upper Paleolithic life and times, suggests that some of the rituals depicted may have to do with puberty. Even as women seem to have initiated young girls with the body emblems and animal forms peculiar to their sex, so men, he says, would early have wanted to initiate boys to the body processes of their sex. Thus the phallic content of at least some of the art. Puberty rites of many people, including the New Guinea tribes earlier described, make such assumptions entirely plausible.

And power restrained

There may be other meanings. Recalling to mind the many shaman figures, one must suppose that the phallic response to religious fervor may also have been associated with images of power, of the feelings men have when at their most male. The ability to call forth at will this emblem of virility must have been seen as proof of superior force, superior command, a kind of conscious transformation, an achievement counter to the accepting passivity of females.

The erect phallus as metaphor of power can be seen among peoples who wear the penis sheath—a tube of plaited leaves or a gourd especially grown and trained for the purpose. The Tchikrin and other Brazilian groups so endow their young men at puberty. "Today you are a man," the new garment announces, but also, "See that you practice restraint!" The penis sheath is thus a sober social directive, not an erotic ornament.

Nor is it so for the Dugum Dani. Donned by a Dani boy at age six, it is ever after the essential article of dress. Each Dani has an elaborate wardrobe of these gourds in different shapes and sizes (older men affect gourds with ends tightly curled). All, however, are so anchored by strings as to remain nearly vertical, presenting thus the very model of the ithyphallic male triumphantly, if artificially, at his most powerful and most controlled.

"Tell me," asks Shakespeare in one of his poems,

> Tell me where is fancy bred?
> Or in the heart or in the head?
> How begot, how nourished?

If by "heart" the poet had a somewhat hormonal implication in view, then today's biological specialist would most certainly reply, "In the head." The male sexual response has in the course of evolution long since ceased to be a matter of reflex and has come to be associated with mind. This probably occurred at about the time that the evolving female's genital skin ceased to swell at estrus, ceased to telegraph bright, colorful signals. Anatomical signals acquired subtlety as they moved toward the front of the female body. Eventually beauty itself became a cultural artifact. As the human male gained conscious control over his sexual responses, he also lost the penis bone, that convenient rod of stiffening which provides an assist to every primate but man. So in some ways the erect phallus can indeed be seen to represent a triumph of imagination and of will. And over time the stimuli to stir imagination were not exclusively sexual in nature.

Models Freudian and otherwise

There was religion. There was also war. And because of that the erect phallus, like the ax, the spear, and the arrow, may have been itself a metaphor of combat. This may explain the fact that phallic trophies have been as popular in some quarters as hands and heads. Swedish rock art of the Bronze Age presents its warrior figures in battle contexts, not only armed but ithyphallic. Ronald Berndt reports analogies among the four east New Guinea peoples he studied. Tribesmen there say quite candidly that "a strong penis makes a strong warrior" and justify in this way the pain of penis bleeding. One of their war songs tells it all:

> Our penes are erect like our arrows . . .
> We go to hide in ambush.

This particular imagery carries over into the sexual act, which is seen by them as yet another kind of armed combat. Love play includes on the part of both partners beating, biting, scratching, even burning. And the male prides himself on wielding his penis like an arrow.

The equation of sex and violence is not foreign to our world. Psychologists know that it is not only lust but also rage and vengeance that trigger rape. They know, too, that sexual dysfunctions depend as often on lack of confidence as on lack of drive. The nonsexual aspects of such behaviors, however, have been largely overshadowed by the Freudian world view which has dominated our culture in recent years. It has triumphed not only because of the persuasiveness of its images but because these images have appeared in the right time, in the right place, and with the right appeal, just when prudery gave way to newer sexual attitudes. The images of Freudian theory are body images, body parts with sexual connotations. By way of these images, personality types have been described as anal, oral, or genital, in the same way that a confident science once characterized the irascible person as choleric, overloaded with yellow bile. Women driven by ambition and eager to succeed in a man's world are said to be suffering from penis envy. In the writing of some Freudians, indeed, the power allusion yields to a rather more literal image: one of women seemingly yearning to strap on artificial phalluses, as early Greek comedians used to do.

Freudian theory has tried to explain in the light of body metaphors all sorts of behaviors and events, from the development of the child to the evolution of culture. All sorts of phenomena not overtly sexual in content have been plumbed for erotic implications. So it is no surprise that the art of prehistory, especially the obviously phallic art, has attracted the obvious explanations.

Yet the phallic response depicted in the rock art continues stubbornly to speak of stimuli in addition to the sexual ones, and of other metaphors. The range of ordinary common sense includes this understanding. Nearly everyone knows of certain homely experiences which seldom appear in the literature of psychology but which form the grist of anecdote. There is, for example, the story of the young married couple, chronically low on funds. They speak wistfully of the chicken sandwiches for which a neighboring delicatessen is justifiably famous. They dwell on the rich house mayonnaise, the lettuce bright and crisp, the chicken, sliced thin but never dry. Suddenly they notice that the young husband has experienced a decidedly phallic response.

Should this be called an erotic act? An incipient sexual perversion? Is it a symbol of fertility, somehow?

One might better shrug and conclude, in today's idiom of tolerance, "Whatever turns you on . . ."

7

DEAD ENDS

The Myth and Magic of Hair

And Samson, the consecrated one, told Delilah all his heart. He said:

> There hath not come a razor upon mine head; for I have been a Nazarite unto God from my mother's womb: if I be shaven, then my strength will go from me, and I shall become weak, and be like any other man.
>
> (Judges 16:17)

Not quite a hundred years ago in the Ojibwa hunting lands around Lake Superior, a lady missionary traveled her pious rounds. Entering a familiar wigwam, she saw a young brave at his sister's knees, being combed and adorned for the day. His hair was very long and very lustrous, and she said to him (playfully, she thought), "Won't you give me some of your hair for myself?"

Solemnly he looked at her and replied, "That I cannot do. Were I to cut my hair, I should lose all my strength."

This tale told by Frances Densmore, observer of the Algonquian-speaking Ojibwa (or Chippewa) in 1905, is not an example of a biblical story well learned and well recited. It affirms a conviction held dear by much of humankind. In Bible lands, in tribal lands, from the ancient Greek to the modern Sikh, hair has been the warrior's strength and the warrior's pride. The Jivaro *tsantsa* is mostly hair, for its original owner, locating there his vital powers, had never suffered it to be shorn. The hair of a Jivaro woman, being an emblem of small account, was generally worn at shoulder length. But a man's long hair, like a man's head, was a visual taunt to the enemy: "Come get it if you can!"

Soul and substance

All sorts of warriors have tried, many choosing the hair of the head, instead of the head itself, for trophy. North American Indians, even the peaceable Pueblo folk, have taken scalps. It was a custom which, as far back as the thirteenth century, left unmistakable traces on the skulls of captives sacrificed by early Iroquois. And in the hair of Basket Maker mummies desiccated fifteen hundred years ago in the harsh Southwestern desert, we find the plaited scalp lock identical to that worn by latter-day descendants and doubtless carrying the selfsame taunt, "Get it if you can!" Scythians of Central Asia took enemy scalps, using them later as table napkins. It is said that the Romans in Judea collected scalps of Hebrew warriors. During the French and Indian Wars, both the British and their adversaries paid bounties for enemy hair, seeing in it not so much an accrual of strength as a tally of ever-weakening numbers.

Toradja of the Celebes to this day bind their hair tightly when they need to concentrate powers and conserve their strength. Loosening hair, they say, will send strength outward to the crops or to rain clouds. Combing is thought to be weakening, and cutting is definitely the wrong prescription for the sick, the pregnant, or for anyone at all while crops are growing. When horse racing came to their island along with the Dutch, the Toradja found yet another application for magical hair. Owners of balky nags, unlikely to win, contrived to steal hair from speedy contenders, thinking thus to equalize the odds.

According to the Maori myth told by Dr. Elsdon Best, one could paralyze a pursuing monster (just what sort of monster we are not told) by throwing in his path a hair of one's head or body. Thrown on the ground, hair would summon forth a gushing spring. And why not? For the Maori, hair was a symbol of the god.

"Rather, a symbol of the soul," would say members of the Fox tribe, originally of eastern Wisconsin. Indeed, one part of the hair—that lying directly under the scalp lock—was thought to be the soul's own home, and the silver comb lodged there was visible proof of contact.

"A second soul," said the Chinese of early dynastic times. Babies, they believed, all hairless and red, brought into this world the soul of blood conferred by their mothers. The soul of breath, the higher soul, was the soul of hair, for hair and breath shared many of the same attributes. The higher soul was thought to appear with the first formal arrangement of Baby's hair and with his first formal presentation to Father, who taught Baby to laugh. Bound to Mother by the belly, the child was bound to Father by the hair.

"Hair," say the Hopi of our American Southwest, "is to wash." Every Hopi ritual is solemnized by shampoo. The head of a newborn babe is instantly sudsed. The Kachina initiation, the boy's tribal initiation, initiation into a ritual society—all culminate in shampoo. At the time of Soyala, the solemn rite that marks the New Year, everybody throws out the old with the shampoo water. Bride and groom tie the knot in hair. Their locks are tangled together in the washbasin while the boy's paternal aunts playfully push their own heads between those of the happy pair. Even enemy scalps in the old days were washed and, with typical Hopi gentility, daily fed.

The Hopi cleansing agent was traditionally the sap of the yucca plant worked into a froth. Now that the Hopi have finally adopted commercial preparations, we find on the market a new shampoo called Yucca Dew. Whether this has achieved popularity in Hopi-land is not known.

In older times the hair shorn from young Hopi children was used in plumping out the braids of newly married women and in certain ceremonial ropes and belts. Long before the heyday of pueblo-building, when Basket Maker ancestors of the present residents lived in the region, hair was one of the few materials available for small textiles. We find hair-weaving still current among Seri Indians on the arid west coast of Mexico. They do not cut their hair for the loom but instead wind the combings in long skeins. Their uses of hair combine ceremony and practicality, the sacred and the profane.

So do those of Aborigines of the great Australian desert. Doughnuts of hair string are worn by women to steady loads borne on their heads. But hair string is also essential in initiation rituals and binds brothers-in-law with a series of exchanges that carry through life. Balls of hair string are pressed against the bellies of mourners to comfort them in their loss and to begin the widow's year of utter silence. String made from the dead man's hair circulates among his maternal kin, enjoining them to vengeance. And in some areas, the identity of the sorcerer responsible for the death is learned by pulling at the dead man's scalp, hair by hair.

Translations of meaning

"Look for the personal, the unconscious meanings," say the psychoanalysts, noting the near-universal association of hair with strength, with sexual potency, with soul. They note, too, that in very many societies a surviving spouse is shorn of hair, even shaved, as a mark of grief. And they conclude that hair serves the individual human being as a displacement symbol, a surrogate for sexual and excretory desires forbidden by the social rules. Hair clippings equal excrement. Long hair equals sexual pride and the unleashing of aggressive drives. To cut the hair, and especially to shave it, is therefore in the analyst's view symbolic castration for the man and symbolic sterility for the woman.

"It is true," replies the anthropologist, "that long hair can symbolize sexual potency, but not every people insists on growing hair with that end in view. Not every people—even the unsophisticated tribal folk—tolerates long hair. And though the shaven head may indeed announce celibacy, it is not always so. Let's look at the variations."

There is no better place to begin than with the Tchikrin, that Brazilian forest folk whose headmen were able to interpret so sensitively and objectively what Terence Turner observed about them. Two groups of Tchikrin have long hair: mature adults and babies prior to weaning. If long hair connotes sexual powers at full throttle, what can these two groups have in common? Not, it would seem, the notion of potency. But, says Turner, that of biological communion comes close to the mark. Babies share this communion with both parents by way of nourishing semen before birth and by milk afterward. A haircut symbolizes the end of this communion, and hair is not again allowed to grow until the young man receives his penis sheath and the young woman bears her first child. The loss of a spouse but also that of a child or a sibling is announced by the shaven head. It is the rupture of biological ties that is thus symbolized, not castration.

Why, if hair equals potency, are so many puberty initiates in various cultural contexts shorn or shaven *after* the ritual that has conferred upon them adult status with all the rights and privileges thereof? Why are mothers in so many cultures shaved after having borne a child? And why the frequent shaving of mourners who could have had no sexual connection with the deceased? Among Walbiri of the Australian desert, it is the dead man's wife, his mother, and his mother's kin whose heads are sheared and gashed. That makes biological sense. But his mother-in-law as well? For indeed she is another principal lady in the drama of grief. Wife, mother, and mother-in-law must not only endure the mutilations of head and hair, they must also singe their pubic hair, surely an even more explicit reference to a sexual bond, which in the case of the mother-in-law is circuitous at best.

In the Trobriand Islands east of New Guinea, a man's death signals a flurry of shaving, not by his maternal kin but by his wife's relatives, who are certainly distant in more ways than one. Perhaps, suggests anthropologist Edmund Leach, the meaning has less to do with biology than with the Trobriand conception of the nature of evil. Trobriand in-laws are often suspected of complicity in a death and so, over the years, have customarily made a greater show of grief than might otherwise be expected. "I'm guiltless; I'm sorry; *truly* I am" is the message of the shining pate.

The shaven head in Trobriand society is against the norm. Indeed, hair and lots of it is the object of warm appreciation. When happy years have passed without a death and everyone's hair has grown long and full, great carnivals of display are held. The point Dr. Leach wants to make is that any variation from the norm can symbolize many things, all having this in common: transition, change. When a person moves from one life position to another, the event can be marked, and often is, by a change in coiffure. When a person is separated, set apart, made different from his fellows, the announcement is made in hair. Whether this involves a cut or a shave or a different style, whether it means combing or not combing, being neat or cultivating disarray, it is the difference, not the kind of difference, that counts.

Psychoanalysis, he goes on to suggest in his article "Magical Hair," locates the origin of symbols in the inherent processes of the individual. The anthropologist believes that society confers meaning, that meaning grows and evolves and is elaborated, but always in specific social settings. "Ritually powerful hair [adds Dr. Leach] is full of magical potency, not because it is hair, but because of the ritual context of its source. . . . It is the ritual situation which makes hair powerful, not the hair which makes ritual powerful."

The nature of hair

Yet hair has not become magical by accident. And that it has indeed been thought magical in so many places and among so many peoples does suggest a beginning in the nature of hair itself. Though only a collection of dead ends, as any physiologist will attest, hair grows. At least, the hair of the *head* grows. Uncut and well oiled, it can, over a lifetime, grow longer than its proud proprietor is tall. Cut it and it grows back to its original length with no visible human assistance. Like the rain, it falls unbidden. Like the plants, it rises unbidden. Like the leaves, it turns color with age. Surely the dawning recognition of this strange vitality must have summoned meaning. Just as human groups found surrogates in the substances of earth for body fluids, for their colors, and for their social meanings, just so did the reality of growing hair serve to make meaningful certain social events and transitions in human life.

More permanent than paint, less final than disfigurement of flesh and bone, alterations of the hair may well have signaled changes that were time-bound and time-limited: the period of mourning, the bridge between youth and maturity, the division of sacred and profane—all journeys completed when the hair has reappeared or been rearranged. Hair cut and encouraged to grow also encourages, by the magic of similarity, the growth of crops and the burgeoning of the child. We can see this among the Toradja and among the Hopi, whose body growth was thought to follow the reappearance of shorn locks.

How humankind came to acquire in the first place this ever-lengthening thatch, no specialist can say for certain. Perhaps, it has been suggested, hair offered protection for the head, ever more full of complicated thoughts. Whatever the reason, it is easy to see that what achieved luxuriant growth on top soon languished elsewhere. While we are not bald all over, good crops of body hair are generally confined to the armpits and pubic region. Everywhere else—given sexual and individual variation—the hairs are small and thin and short.

Bodily sparsity, being a corollary of abundant sweat glands, is more easily explained than cranial luxuriance. Evolving humankind, it now begins to appear, made a living by hunting in the daytime, by tracking down and walking (not running) down the large animals that were its prey. Such exertions required and eventually selected an efficient cooling system, one that worked best in the relative absence of hair.

The zoologist William Etkin years ago proposed yet another explanation for our loss of body hair. As all the world knows, he pointed out, primates do a lot of mutual scratching. Socializing for them means cuddling up to groom. When humankind acquired speech, he added, we found yet another way to get our strokes, and hair went out the window. And if you've ever stood back and listened to what passes for conversation at a suburban cocktail party, you'll tag it instantly for what it is: verbal grooming.

Dealing with the pelt

Unlike the hair on the head, body hair reaches a prescribed length and grows no more or breaks off. The hair of animals reaches a prescribed length and grows no more. This may explain why humans began initially to pluck and scrape away the body foliage: they found the animal comparison all too demeaning. The Yahgan in Tierra del Fuego made the connection explicit. Said one man to a poorly turned-out friend, "What! Unplucked! Do you want to look like a dog?"

Whether or not contemporary depilators have this particular image in mind, the removal of body hair has, in many cultures, been considered essential for proper appearance or to protect aesthetic sensibilities. No self-respecting person in such groups would dream of remaining as Nature made him. The old might let themselves go, but not the person of middle years. The hair of arms, chest, legs, pubic region, nose; the eyebrows, lashes, whiskers—all or some have been enthusiastically removed according to patterns prescribed in various settings for the two sexes. Removal of pubic hair has been a must among virtually all tribes in the Americas. Hopi damsels are said to have distributed these as love tokens. And Cubeo women, says Irving Goldman, continue to pluck, even though the region is now hidden under skirts. Modern American women, who insist on shaving the hair of calf and armpit, would consider the removal of genital growth only for an operation or in expectation of performing a striptease.

Before the invention of tweezers and steel razors, depilation was achieved by use of various devices: box shells, bamboo splints, wax applied to thumb, scrapers of obsidian. Maya mothers are said to have scalded their babies to prevent the growth of body hair. And among the Tchikrin, biting off the lashes of one's sweetheart is considered a particularly charming bit of love-play.

Although doodled human forms on Paleolithic cave walls depict the first Europeans as mainly clean-shaven, the most enthusiastic depilators of historic times have been peoples naturally less disposed to body hair. The Africans and Asians of Old World and New possess the richest lore focused on the negative meanings of body hair. Australoid and Caucasian peoples, on the other hand, realizing that effective depilation might consume a good part of every waking day, resolved early on to make capital of what came naturally. European Jews of medieval times thought body hair lucky. The Ainu, an ancient Caucasian people of northern Japan (the so-called hairy Ainu) have, in other days, gloried in their pelts and even tattooed on their women's faces the semblance of a blue mustache. Elsewhere the naturally hairy-lipped woman might be considered extra-passionate or a possible witch.

Naturally hairy groups are apt to attach great significance to beards. Moslems swear by them. Uncles in some Italian dialects are named for them. The various Anabaptist sects of North America, still frozen in Reformation times, identify the married man by his growth of beard. Even enthusiastic depilators have seemed to find a few chin hairs impressive, seeing in them a mark of wisdom or authority. Egyptians, by Middle Kingdom times (2000 B.C.) utterly clean-shaven from top to toe, still retained the goatee interwoven with gold threads as part of the pharaoh's regalia. Some specialists have seen in this vestige, as in the ceremonial crook, an apotheosis of an ancient nomadic goat-herding existence eventually replaced by farming ways as emblemized in the pharaoh's agricultural flail. In any case, the pharaoh soon came to depend on a false "chin-piece" bound to the head with various strings and bands. One female pharoah, the famous Hatshepsut, was depicted wearing the goatee.

Management and disposal

That which grows must be kept within bounds lest the work of the hands be jeopardized. Beards must be rolled or trimmed. Hair must be bound or cut. If cut, what to do with the clippings? While it is true that some peoples have considered them unclean, on a par with excrement, others have offered shorn hair to the gods, depositing it on the nearest shrine. Polynesians in general did so. Whether unclean or sacred, hair and nails of fingers and toes (also growth industries) have been accounted a continuing part of their proprietor and have been everywhere given careful disposition. In the wrong hands, after all, they could be used to work evil. So, depending on the place and the people, clippings have been buried, burned, thrown into the bush or down crevices, embedded in cow dung, or deposited in special containers to be buried with the owner at his death.

In arranging, cutting, and depilation, women have (until very recently) most frequently served as beauticians. The sister cut her brother's hair, the wife her husband's. In the few known cases where the reverse has been true, always onlookers have marveled at the true love—if not downright silly infatuation—on display. Until historic times and to this day among tribal folk, hair has been a man's crowning glory, to be kept long and intricately dressed. It is most often in complex, literate societies that women have come to monopolize the art of the hairdresser, as they have that of the cosmetician. It is in historic times that women began to wear their hair longer, and in historic times that women's hair came to be regarded as an erotic organ. So spellbinding were a woman's tresses that, from earliest times, pious Jewish women were adjured to keep them covered after marriage. Eventually, women attributed family success to their diligent observation of this rule. "The beams of my house have never seen my hair," boasted one.

By the Middle Ages, Jewish matrons were disposing of hair altogether, shaving it off and covering baldness with a kerchief. Later, wigs were added under wraps. The kerchief then shrank to token size, and the wig came fully into view. As late as 1901, immigrants from Central Europe were coming to these shores wearing their wigs. Then, stung by the mockery of the natives, they decided to let their own hair grow, adapting it to the style currently in favor.

Occupational styles

In earlier times, hairstyles were intended, as Dr. Leach has pointed out, to announce a status, a position, or to signify its change. Holy persons have always been set apart by lots of hair or the lack of it. One thinks of Sumerian or Egyptian priests, their shining heads like peeled eggs. Indeed, the Sumerians maintained ritual nakedness (and consequent body hairlessness) when serving the gods, a fact which may illumine the ithyphallic nudity of some human figures in Ice Age caves.

Buddhist nuns and Christian nuns and Aztec temple virgins alike had pates shaven or shorn. So did Buddhist monks and Tibetan lamas, while their Christian counterparts were required merely to shave their faces and the crowns of their heads. In older times, the pious Hindu gentleman was shaven except for an obligatory cranial tuft which was the mark of his religious affiliation and remains so to this day among the American devotees of Krishna.

Some specialists have thought that the shaving of holy persons connotes celibacy or, as in the case of the Hindu layman, a sex life with restrictions. If so, it is a dictum unknown to Coptic monks in Ethiopia, marooned on inaccessible peaks and plateaus. No female being, not even a hen, can be hauled up to these bearded ascetics in their aeries. The same stricture obtains for monks of the Eastern Orthodox rite. And they are as bearded as the secular priests, for whose orders marriage is permitted. Among modern Greeks, the beard is standard uniform for holy men. The not-so-holy are expected to grow only mustaches, the hallmark of virility and male attractiveness.

Hebrew priests of biblical times could not dispense with their beards or long locks, though these were carefully trimmed. Only for rituals staged under the most awesome and terrible circumstances was every hair removed. The Nazarites, those consecrated to God, allowed no knife or razor near unless, like gullible Samson, they entertained a weakness for conniving females.

Shamans everywhere are expected to have a better-than-normal stand of hair and to leave it uncut. The Aztec priest, according to Díaz del Castillo, had long hair uncombed and so matted with blood that he was a stink even unto nostrils habituated to the filth of Spanish cities. Sadhus, those wandering ascetics of India who not only renounce the world but ignore it altogether, cultivate to this day hair just as matted but laced with ashes instead of blood.

Style, gender, and punishment

Sex differences are nearly always announced in the hair, though there are a few regions, such as the Orinoco–Amazon Basin area, where men and women affect an identical cut—the stylish bowl. (During the unisex sixties, identical manes were here prescribed.) Each age group ordinarily has its style, the most effort often being expended on the maiden or youth ready for marriage. Even Eskimo men, reported Peter Freuchen, under wraps as they are, appreciate a young girl with lots of hair. He tells of one, a baldish and thus unmarriageable damsel whose exasperated father finally fitted her out with a musk-ox's hair in order to attract a prospective bridegroom. Success at last! Until the falsies were discovered.

Sixty years or so ago, young American women announced their eligibility for marriage by putting up their hair. Thereafter, the hair was loosened only in the most intimate and/or erotic circumstances. This custom so marked the idiom that we still whisper to a group of friends, "I'm going to let my hair down," when we are intent on utter candor.

Special styles for parents (especially mothers) are often prescribed by a culture. So are styles for widows and widowers (we spoke of this earlier) and styles for the old (usually slovenly and in disarray). Jobs and special statuses, if these exist in addition to the shaman's position, are often announced in the hair. And the sinner, the criminal, the degraded one are likewise made known. The Hopi and the Chinese of Han Dynasty times punished theft with a loss of hair. Many American Indian tribes stigmatized the adulteress with shorn locks or a shorn nose. Perhaps the one was meant as warning, the other for the persistent recidivist. The erring Hindu woman, the Teutonic woman, and the Jewish adulteress were shorn of hair but kept noses intact. After World War II, people who had been oppressed by forces of occupation marked female collaborators by shaving their heads. Perhaps the saddest shaving of all is that given to the traditional Hindu widow, for it is meant as a kind of punishment. Had she not sinned in an earlier existence, would she have lost her husband in this life? And the younger the widow, the more heinous that forgotten crime is presumed to have been.

Soul mates

Some styles are oddly convergent. The monk's tonsure, symbol of Christ's crown of thorns, is also to be found among Indians of Venezuela and northern Brazil. One of these, the Yąnomamö, say the shaven pate marks the spot where blows of a rival's club are meant to fall. A man has the scars to prove it. His wife also flaunts her scars, administered not by a rival but by her husband, who thus shows sufficient attentiveness but stops just short of outright murder.

The roach, that narrow stand of hair which crowned the otherwise shaven head of many a native American warrior, finds repetition in the soldier's helmet of Classical Greece. It may have derived in turn from much earlier folk on the Hellenic mainland. We see the heads of such warriors inscribed on the Phaistos Disk, that undecipherable document found on the island of Crete and dating perhaps to 1700 B.C. or thereabouts. In Egyptian paintings of about 1100 B.C., invaders wearing such headdresses are identified as People of the Sea.

The Dugum Dani and other tribesmen of West Irian, Indonesian New Guinea, wear their very curly hair in what resembles nothing so much as an old-fashioned dust mop. Each strand is carefully twisted, like a thread of yarn, to hang ear-length on the sides of the head, just to the brows in front. That selfsame style appears as well among contemporary African groups such as the

Swazi of South African Swaziland, who give the coiffure a reddish tint with liquefied hematite. It is a style once more widespread in popularity; that we can see in Egyptian tomb paintings of Africans and, indeed, of Egyptians themselves. This particular convergence doubtless represents, in the case of the Swazi and the Dani, a natural elaboration of hair similarly textured and curled. Whether the more southerly African styles represent an imitation of something created by their fashionable northern neighbors we cannot say. Since the Egyptians affected wigs, the imitation may well have been quite the other way around.

Substitute hair and hairy hats

The fondness for wigs appears and disappears throughout human history and has, it would appear, a more than casual connection with that bane of existence, the head louse. Maybe not bane altogether. The Toradja considered the comb on which the louse was introduced a gift of the gods, for delousing, in their view, along with joint combing and tending, was one of life's sensual pleasures. Delousing is standard grooming fare in every tribal society and was in this one until not so long ago. The word *fare* is used advisedly. Delousing often provided the operator with several nibbles of protein. Glimpsing the congenial tribal group chatting and delousing, one wonders whether grooming did indeed go out with body hair, as Dr. Etkin suggests.

The wig over a shaven pate is not the only possible response to an infestation of head lice. Mohave and Maricopa Indians of Arizona are said to have packed their hair into mud coiffures. When the mud was removed, the lice were dead, and the hair was left in shiny-bright condition. Perhaps some origin of this sort accounts for the popularity of urucu-plastered locks in South America and mud coiffures in East Africa. Moving among cattle-herding Nilotes, from the Sudan southward to Tanzania, one can see warriors wearing what appear to be fanciful helmets. They are actually molded from mud, cattle dung, and the warrior's own hair. Such arrangements take days to construct and are often gorgeously colored and adorned with plumes. Among the Jie, the Suk, and the Turkana, the helmet is shaped into what resembles a large chignon. To the original "do" a man will add those of his deceased father and of all the other men he calls "father" until he has a heavy rope of buns extending down his back.

Lice have had some effect on American hairstyles. When, during World War I, it was learned that the louse carries typhus, soldiers were made to cut their hair very short, something very close to the crew cut that remained popular right through World War II.

Hair and politics

Hair, like clothing, has always been the target of sumptuary laws. Beards on, beards off. Hair long, hair cut. And cut always according to certain supposed realities of the universe. When Christianity extended among the bearded Teutons and Celts of Europe, the clergy were ordered to shave. Bearded Greeks, on orders of Macedonian Alexander, were shaved clean lest they present a handle to their enemies. When beardless Japanese conquered the hairy Ainu, the latter were ordered to shave. When these same Ainu fled to the Russian Kuriles and encountered Russian Orthodox clergy, again they were ordered to shed their hair.

Manchu conquerors of China ordered their subjects to shave all but a single long plait, or queue, and many a mandarin preferred rather to forfeit his head. It appears that the rule was devised as an instrument to curb dissent. The Chinese ready to desert could not do so without first shaving his queue and waiting for the hair to reappear. Meanwhile, he could be apprehended and imprisoned. Later, when the queue grew again, it provided a convenient lasso for the executioner's assistant.

American hair

"The more things change, the more they stay the same." So the saying goes. One might better add, "And the faster things change, the more they return to what they were in the beginning." The woman who has been around for fifty years cannot but be bemused by the rage in the seventies for frizzy permanents. In the early years of the permanent wave, frizz was the mark of the cheap beauty parlor or a hair set done at home. Now the stylist with every improvement of modern chemistry at his disposal declares his admiration for the "new" frizzy top and revives beauty technology of the twenties.

Fifty years ago it was the daring girl who cut her hair. Ten years ago, the same girl was daring again. Long hair, pressed straight on an ironing board, had become a female badge of nonconformity. Or rather, since Americans prefer to nonconform in groups, it had become the uniform of an age set wending its way adultward.

Fifty years ago the man with long hair was sure to be a classical musician, probably (so the average Middle Westerner believed) a conductor living in New York, certainly a "bohemian" with all the stylish unconventionality that the term implied. Ten years ago the typical long-hair was a young man whose sole connection with music was by way of the guitar; who, together with his peers, constituted a new class of world nomads but who most often took "trips" without going anywhere in particular. Frequently long hair was deliberately unkempt, for like the religious disarray of the sadhu, it was meant to signify a certain rejection of the established order, of its sexual strictures, its work ethic, and its fondness for the material trappings of success. Renunciation, however, was not total. The new asceticism permitted ownership of the indispensable goods of life—of the stereos, transistor radios, and cars whenever fond parents could be persuaded to make these available. Oldsters, finding rejection painful, inveighed so bitterly against the revolution of hair that a popular musical was created to make their discomfiture complete.

Now the comrades-in-hair are for the most part pushing thirty. Matted locks have undergone elegant styling by the new breed of barbers who create individual "looks" in hair, longish but neatly trimmed and worth a king's ransom to achieve. Certain styles once again reflect certain personalities and even certain jobs. There is the style that marks the youngish executive, stable yet receptive to new ideas, energetic yet mature, definitely on the way up. There is a style to suit the older man, hiding his paunch and eager to "swing." There is a style for the bookish and a style for the athletic. Here one finds interesting variations on a theme. Baseball players, football players, and long-distance runners are all identifiable by their distinctive cuts and growths of hair.

Females just graduating from the Age of Aquarius tend to cut their manes. Short, straight styles executed with a battery of new gadgets permit a windblown, casual look, with plenty of volume and a minimum of split ends. Alternatively, they are reverting to older styles with lots of puffs, curls, and elegant chignons. Some aging holdouts retain the long, straight tresses, unbound and falling in the eyes. No scissors, thank you, and no hairpins, either. Many admit to having nightmares about scissors and the loss of hair. Such dreams provide analysts with plenty of support for their theory of castration anxiety. Yet it must be remembered that something long grown and long tended is never lost without regret. Before violent exercise became an article of faith, even the man with the paunch regretted a loss of girth. Patting his tummy, he could gloat, "I've got a big investment there!" And so with outsize hair. It represents an investment of time.

And the absence thereof

Still, with time, something cut can be regained. The long or the short of hair is, for women, a matter of decision. For some men, there are no decisions to be made. There is only the law of diminishing returns and the looming inevitability of baldness. Typically, this is a condition of hirsute peoples and is virtually unknown among Asians of Old World or New. So rare, indeed, is baldness among the Eskimo of Canada's Baffin Island that all such men are said to be descended from a famous shaman who, long ago in famine times, led his people to a better land. All through the dark journey they followed the beacon of magical flames issuing from his head.

We can see the shaven equivalent of a receding hairline among Tchikrin adults, both male and female. Presumably the pate is exposed to receive the red paint that both signifies and guarantees vitality and dash. Japanese samurai also favored the exaggeratedly deepened brow, though more, it would seem, as a fashion expedient than as a philosophical statement.

What is done artificially to complete and enhance the appearance always seems to be more acceptable than the natural state of things. And baldness, especially baldness early in life, has not yet acquired positive symbolic content. The gentleman so afflicted has traditionally covered cranial nakedness with headdresses, other people's hair, or his own side locks grown long and combed over the top. The balding Julius Caesar, it is said, never removed his victor's wreath of laurel leaves, whether indoors, outdoors, or in milady's chamber.

The Hebrews of biblical times mocked baldness, but later, when scholarship came to be held in high regard, attributed loss of hair to much weighty thought and diligent study. Activity in the upper story, they felt, left no peace for the growth of foliage on the roof. Over the years, that particular myth seems to have made sense to increasing numbers of people. How, otherwise, would we have received our current image of the extraterrestrial—denizen of the galaxy and saucer pilot extraordinaire? Typically he or she displays an outsize head, bulging with weighty thoughts and as innocent of hair as a billiard ball. Indeed, the prophets of our own future tend to see us with much the same sort of appearance. All-over baldness thus bids fair to betoken superior intelligence and a presumably higher rung on the evolutionary ladder.

Yet another myth has led many a hairless hopeful to declare, "Baldies are more virile," while generously offering to provide instant proof. Perhaps to emphasize the point, there are those who have made capital out of a dwindling asset by shaving off whatever fringe remains. If enough men blessed with the perfectly shaped head and personal élan of a Yul Brynner decide to go all the way, a new style may be on the horizon. And more than that. If the two equations—baldness equals brilliance, and baldness equals virility—can be merged, we may experience a metaphorical revolution in the body suite. The vital associations of hair and its millennial imagery—erotic hair, magical hair, and hair of potent strength—may soon be as defunct as the dodo.

Out, Samson and the Indian brave. Farewell, toupees, implants, and scalp massage. Away with blow dryer and with brush.

Razor, hello.

8

OF TIME AND BODY

The Cosmic Connection

"If only I'd had bosoms, I could have ruled the world!" The remark is credited to Julie Harris, the fine American actress. Doubtless she had in mind an audience reaction that included quickened pulses as well as cerebral satisfaction. And she was right about what would do the trick. Since fashion decided once and for all to bring the female leg into permanent view, no other part of the anatomy has attracted quite so much erotic attention as the breast. It has been bound and unbound; cupped, separated, and cross-your-hearted. It has been alternately cantilevered and encouraged to hang free. The once-invisible nipple is now allowed definition. Torso garments have been sweater-tight to reveal the silhouette, loose and clinging to offer tantalizing peeks. Though the décolletage has not as yet arrived at Cretan latitudes, it continues to plunge.

What we tend to forget is the fact that the breast has function as well as form. It is the symbol of the mammalian rise in the world, the evolutionary prominence achieved by way of tender, loving care (plus a few cortical renovations). The mammal mother's milk is designed to be the perfect food for her own baby. This is true of the platypus, whose milk exudes from leathery pores without benefit of nipples (doubtless proving that function precedes form). It is equally true of the human species. But technological advances give our mothers a choice in the matter. To lactate or not to lactate becomes less a matter of health than one of convenience or aesthetics.

It was not always so. In older, harsher times, among folk who lived close to the land, the baby whose mother went dry soon wasted away. The baby whose mother died in his delivery accompanied her, living, into the grave. And the baby born out of season, born before the preceding child could yet be weaned, was slain immediately on emergence. For milk was life, and the rules of life were stern and hard: first come, first served.

The talismans of life

In 1910, Sir Arthur Evans, Keeper of the Ashmolean Museum at Oxford, traveled on donkeyback through Crete. He was in search of seal stones carved long ago with ancient writing and traced to that island. He found them threaded on strings and tied around the necks of Greek women, all nursing mothers. When he tried to buy the stones, they recoiled. The talismans, they told him, were "milk stones" necessary to the continued nourishment of their children, impossible to relinquish, however dearly he was willing to pay.

How Minoan seals, relics of a bygone civilization, cylinders meant to be rolled as signatures over wet clay—how these seals became milk amulets we cannot know. But because of them, certain tiny pendants carved during the Ice Age begin to make sense. Fashioned of ivory, they are nothing more than complex beads, beads that represent, each of them, a pair of swollen breasts. Such strings of beads were in continuous use for over twenty thousand years. They have turned up in Israel in sites once occupied by Mesolithic folk. They have turned up in the

Greek islands and in Switzerland at a time coincident with the beginnings of horticulture. They have turned up in France and Italy, where men once built mighty monuments of huge stones in megalithic celebration of celestial events: the movements of sun and moon, the changing alignment of star clusters. While men looked up, women looked down on their nursing infants, secure in the knowledge that the milk would not fail and that each loved and precious life would go on.

Among hunting people everywhere, breast-feeding is a lengthy procedure, ideally continued until the child is four years old or older and, taunted by his playmates, elects to wean himself. Prolonged periods of suckling doubtless marked the life cycle in Ice Age times as well. Long dependence on the breast as font of nourishment brought into being symbolic associations wholly missing today. The poses of female figurines carved during these times, arms draped over or cuddled under enormous pendant breasts, were echoed in the Neolithic. Middle Eastern "goddesses" then were invariably shown with hands cupping the breasts in offering. There was surely nothing of the erotic in any of these little figures. The feeling, it seems, was analogous to that which inspired medieval painters to portray the Madonna giving suck to the Child Jesus.

In Çatal Hüyük, you will recall, plaster breasts were modeled around the beaks of vultures, the jaws of the wild boar. Clearly an ambivalent message was here conveyed, an evocation of death in life, of want in plenty, of terror in love, and all expressed by a female metaphor introduced during the Ice Age.

Life and its quickening

Always there has been associated with female processes and parts a certain ambivalence. The cycles of blood and the cycles of birth, the one dreaded by man, the other desired, are each mysterious and unbidden, beyond understanding and therefore beyond control. And how much more awesome both events when men did not yet know the partnership they shared in summoning forth new life. That there was a time of innocence in this respect is suggested in the biological folklore of Australian Aborigines. We explored this earlier. But they were certainly not the only innocents. Malaysian Semang believed (and may still do) that women received baby spirits by eating the flesh of a certain bird. The husband hopeful of issue killed the bird, brought it home, and then preserved its bones until after the blessed event.

Greek myths, says Robert Graves, evoke a time when ancestral spirits were thought to enter a woman's body in a gust of wind. That notion was preserved (at least with respect to horses) far into Roman times, when prized brood mares were regularly turned for impregnation hindquarters to the wind.

Andaman Islanders told Professor Radcliffe-Brown that baby spirits inhabited stones of the surrounding reef and that a woman had only to wade out at low tide to get a pregnancy under way. During World War I, when Bronislaw Malinowski was interned in the Trobriand Islands, he encountered among its Melanesian farming folk an unshakable faith in maternal responsibility for new life. The man's part was confined merely to "opening the way." "There are baby spirits," he was told, "floating out there on driftwood." Hovering shades of Grandfather or Grandmother would kindly deliver them to the proper address. When a Trobriand husband had been long years at sea and returned to find his household newly augmented by offspring produced during his absence, he rejoiced and received them in his arms as gifts made by generous ancestors. The same causes were invoked in the multiplication of pigs. "It must be so," the Trobrianders insisted. "All our male pigs are castrated." Of course they overlooked the eager presence of wild boars lurking in the bush.

Ethnographic literature offers any number of intermediate positions between no understanding of biological paternity and the calm assumption of its preeminence. Ancestors work with Father to ensure conception. The shaman helps. The moon helps. The gods help. Or the time, the hour, the season, the lucky day. Take your pick.

All these suggest a time in antiquity when knowledge of the male responsibility in conception was totally unknown. Just when such a state of mind was general nobody will ever be able to say. It is enough to imagine the awe, the wonder, even the fear that surrounded an event so very precious and so miraculous.

Enter the mother, fair, fat, and fertile

Not surprising, then, to discover among the earliest art forms of the Ice Age, among the elegant and realistic figures of animals carved in bone and later on walls, deeply engraved vulvas and female figurines, sketchy and crude. For the force of life made manifest in women was not *a* woman, not *particular* woman, merely Woman. And so the carvings and sculptures, as they took on style and sophistication, were seldom true to life, were most often rendered in a conventional mode. Faceless and anonymous, with rudimentary arms, legs, and feet, the figures focus on and exaggerate to a grotesque degree the secondary sexual characteristics. Pendulous, ballooning bosoms bulge over a swollen belly and plump vulva, while equally swollen buttocks bring up the rear.

It may be that the lavish display of flesh represents as much ideal as anonymous principle. Present-day hunting-gathering folk are seldom fat. Their diet—rich in protein and fiber, lacking in grain and milk—does not permit it. Indeed, the lack of fat among nomadic Bushmen retards the onset of the menses and inhibits ovulation, thus restricting population increase. So we are told by Gina Bari Kolata. The northern herd animals of Ice Age times may have yielded considerably more blubber than the stringy antelopes of the African Kalahari. But until we can develop a better understanding of the diet, we had better not relegate Ice Age woman to the local fat farm.

Whether actual or idealized, many female figurines of the time bear the polish of long handling and traces of red ocher, the color of life. And they have been found throughout Eurasia, from one end to the other, all of them referable to early cultures of the Ice Age. Toward the end of that period the little "Venuses" were eclipsed by even more schematic forms—the pendant "squats," the hourglass carvings, and disk-shaped vulvas—and would not be resurrected until Neolithic peoples found in them an evocation of the earth.

Whatever the nature of the female principle, man shared in its beneficence and its promise. We see that in the male burials of the late Ice Age. Included with the grave offerings were many sacred images of the life-force, abstract but female.

Sort through records of the world's hunting tribes and you will find echoes of this ancient principle. Even among hunters whose relentless focus on war shoved women into second place, the female principle was accorded signal honors. Warlike Indians of the North American Plains credited culture heroines with tribal beginnings. The Sioux tell of a maiden saved by a giant eagle from a world-destroying flood. Naturally, the eagle is transformed into a Sioux warrior and together they repopulate the world. The Cheyenne Council of Forty-four, the most sacred of the tribe's political structures, was founded by Short Woman, a magical being saved by animals from her father's treachery. The most solemn moment of the Mandan O-kee-pa occurred when a chosen matron, claiming to be the *Father* of Buffaloes, the bearer of life, drove out an evil spirit which threatened both animals and Mandan women.

All across the Eurasian north, the tribes told of the Old Lady Who Began It All. The Lapps called her the creator of the universe. So did the Tungus. The Ket said she shook birds from her sleeves. The Eskimo gave her command of the sea mammals. The Gilyak called her the Horned Mother.

In North Australia, men speak of the Mother who introduced the stone circumcision knife, through the use of which boys are reborn as men. Like human mothers, they say, she prefers her sons, and so women are excluded from her rituals. A similar cult is found among tribes of the Mount Hagen region in eastern New Guinea. There men honor the Female Spirit, who appears to them in dreams, painted all in white and adorned like a bride. Although she represents fertility, she is also pure and thus protects men from the menstrual pollution of their actual human wives.

For Pygmies of eastern Zaire, the Forest is the ultimate reality, the ground of being, the great soul to whom they sing nightly in times of sorrow and trial. It is the men who celebrate the Forest, but now and again women intrude, led always by an ancient crone. Scattering coals of the sacred fire, they bind men to silence with a long vine. And then they retreat, leaving their leader to release her captives. Afterward, each man comes to touch the old woman's hand as if to ask a blessing.

Body seasons and reckonings

Whatever the bounty and benevolence of the female principle, ordinary females must produce the living demonstration thereof. They bear the children and, in order to do so with reasonable convenience and due regard for whatever proprieties obtain, must keep track of the cycles and predict the birth. The normal term of pregnancy, calculated from the mother's last menstruation, is 282 days—the length of ten menstrual cycles of twenty-eight days. No woman today has to trouble herself with such calculations. No woman today has to predict the arrival of the blessed event or mark the passing days. Her doctor does it for her. In other days and other cultural settings, however, this was a wholly female responsibility—even a female secret—of little interest, usually, to menfolk on the scene. And male ethnographers, seldom privy to such information, have included in their life-way descriptions a vague assumption that tribal births were simply reckoned according to the seasons. Some have not troubled to assume. In 1823, Father Ivan Veniamenov, the Russian observer of the Alaskan Aleuts, had this to say: "Women had their own omens and charms for pregnancy, childbirth, and so forth, the details of which I do not know and, it seems to me, are not worth knowing."

Often we learn of record-keeping only when the requirements of ritual are such as to make knowledge fairly general. Among the Manus of the Admiralty Islands off New Guinea (a people brilliantly described by Margaret Mead in 1928), women counted ten "moons" to pregnancy, using bundles of sticks for the tally. An exact date was essential. The expectant mother's brother had to choose the "right" place for the delivery—his own home or the paternal home, in which case the father moved out and was replaced by his wife's relatives, who supervised the birth. They also supervised the solemn round of financial transactions which then ensued between themselves and the baby's father.

The California Yurok (of whom we have heard before) were equally concerned with finances. And they were obsessed with the moral proprieties. The baby born during the "bastard moons," early fall and winter, shamed his parents, was living proof of their lack of moral fiber and self-control. For, said the Yurok, the self-respecting couple (which means the wealth-respecting couple) will not offend the dentalium shells by hanky-panky in the house. They will remain continent until able to sleep out of doors. No wonder, then, that births were carefully calculated. Psychiatrist Erik Erikson said it was a ten-period count for a firstborn. Arnold Pilling in 1968 told Alexander Marshak of a day tally of twenty-eight sticks and a month tally of ten kept in a second basket.

Marshak suggests that such tallies may have been kept by women of Upper Paleolithic times as long ago as thirty-two thousand years. Many if not most of the crude and schematized female figures dating to these early times bear notches or engraved lines arranged in sets. These do not necessarily correspond to a true lunar model. Nor yet does the twenty-eight-day cycle consistently appear. It must be remembered, however, that females are not paragons of physiological consistency. The cycle for some women may complete itself in twenty-one days or run for as many as thirty-five. And if female tallies of the Ice Age represent the crossed-off days of individuals, we must expect some variation.

It is just possible that in these little tallies (if indeed this is what they are), these records of body cycles, of the fearful fluxes of blood and the fruitful absence of same, we can read the origins of record-keeping on a grander scale. Herein may lie the beginning of the calendar and of the many schemes whereby time is ordered, remembered, and managed.

Heavenly bodies

At some point, Alexander Marshak considers, someone may have noticed the similarity of the female and the lunar cycles, or projected from one to the other. The moon, after all, is visible in the sky for a period of twenty-seven or twenty-eight days at a crack. And in very early times, perhaps thirty thousand years ago, what appear to be quite accurate lunar records were being kept. At some point, the two sets of records and their associated images, rites, and metaphors were combined. Perhaps such a sacred marriage is embodied in the lovely "Venus" of the Laussel cave. A bas-relief of seventeen inches, oft rubbed with ocher, this traditional, lavishly fleshed female holds in her outstretched right hand a bison horn.

She is not the only figure in Laussel so to do. Certainly one and possibly another crudely rendered female have horns in hand. Now, to see new and old moon in animal horns and boar tusks (as among the people of Melanesia and New Guinea) is an easy metaphor. Bison horns are, like the moon, unmarked. The horn held by the sovereign lady of the cave, however, bears thirteen clearly incised lines. Thirteen is the year count of the repeating female cycles and of the year's visible crescent moons.

The picture of the hunter anxiously scanning the night sky from the mouth of the family cave is charmingly persuasive. And one cannot deny the impact of heavenly events on the mind of man. (Even chimpanzees are said to sit, now and then, in rapt contemplation of a jungle sunset.) Still, for all the possible advantages in a seasonal tracking by way of the moons, a certain element of sheer intellectual curiosity must be invoked. And while one could not want to deny this sort of curiosity to Ice Age man (or his forebears), one cannot but suspect a more basic impetus, a nudge of a more elemental sort. If first colors were body colors when Nature was ablaze with brighter hues, why not relate first rhythms to body rhythms?

Enter the father, lusty and lean

Remembering the sort of ceremonial tit for tat practiced by many of the world's males with respect to menstruation, one wonders whether lunar reckoning might not have had a similar stimulus:

> Anything you can do, we can do better.
> Any mark you can make, we can make more of.
> Take your tallies; we've got ours!

Such a guess is possible only by use of hindsight, by extrapolation from the myths and lore of people whose life ways appear to suggest earlier usages and older beliefs. Among such people the moon is a male being. The Andaman Islanders call him Father of the Skies, with the sun for wife and all the stars as their children. Others invoke a less paternal image. Anong them, the moon is seen as a mighty hunter and a mighty lover as well. By copulating with women he causes menstruation. "The moon has seen her," say the Jivaro of the girl in first menses. The Cubeo believe that every time is a first time, the hymen magically growing back for a woman's next lunar affair. For some, like the Eskimo, the male moon is responsible not only for menstruation but for reproduction as well—for the fertility of all living things. The female sun has nothing to do with that.

If records and powers of the female cycle were anciently coopted, it was not an easy victory or one achieved without guilt on the part of the new owners. World mythology is replete with tales of theft and trickery, and always at female expense. Ituri Pygmies stole fire from women, who, in their turn, had taken it from the great Forest. Australian Aborigines stole rituals, New Guinea people stole the flutes, as you'll recall. So did countless tribes of the Amazon Basin. And all forbade women to see again what once they had owned. Men of West African secret societies, anciently betrayed in war by wives loyal to their natal families, wrested from them the dominance of the masks by subduing a forest spirit more powerful than the female tutelaries.

The Yahgan overturned the entire order of things. Women, they said, had once ruled by the power of witchcraft. Then one fine day, men killed all women over twelve and brought up the remaining girls to respect the proper order of things. And to maintain that order, the Yahgan yearly performed a ritual reenactment of the takeover. Horribly painted and masked, they practiced noisy intimidation. And the women and children obligingly quailed and quavered.

Samuel Johnson once said, "Nature has given women so much power that the law has very wisely given them little." The message of the ancient myths seems essentially the same: "Keep 'em down or they'll take over!"

And yet the battle of the sexes can as easily be read in another light, and probably should. There was no chauvinist plot hatched in time's beginning. Men responded, rather, to a perceived conflict between the natural and the cultural. Never mind what variations were later played on that theme. The initial polarities were surely part of a continuing heroic human effort to constrain the body rhythms, to subject them to the dominion of mind.

From female rhythms and their recording to the heavenly rhythms recorded by men, from the processes of the flesh to those of the skies, from female properties to male designs—it makes a logical sequence. "One starts a journey of a thousand miles," as the Confucian saying goes, "with the first steps." One first comprehends the human body. Heavenly bodies come next.

Keeping up with the times

It may be that this sequence of events obtained wherever early human beings lived and traveled. Maybe not. We do know that the lunar revolutions not only gave the first framework of time but continued to do so long after the disadvantages of this kind of reckoning had become apparent. Some lingering sacredness in the old scheme of things seems always to have prevailed.

Lunar calendars and a lunar mythology preceded solar and stellar reckoning in Egypt. In Sumerian mythology, Utu, the sun god, is the son of the moon god and brother to the goddess of love. A celestial imprimatur to the temporal ordering of things. Long after the Maya had devised at least three different sorts of calendrical systems (using all concurrently), they continued to complete each engraved dating of historical events with a number of glyphs describing the current position and phase of the moon. Three thousand years ago the people of Southern Arabia honored a preeminent moon god always represented by the head of a bull. (Archaeological excavations of the 1960s have revealed the connection.) At the time of Mohammed, reckoning by the moon still prevailed in the region. And to this day the official Islamic calendar continues to be lunar, according to the Prophet's decree.

The drawbacks in lunar reckoning are obvious. The twenty-seven or twenty-eight nightly appearances of the moon do not immediately repeat themselves, as do the twenty-eight days of the female cycle, one after another for thirteen periods of the year. Always there are the one or two days during which the moon is altogether dark. If one keys the thirteen visibilities of the lunar year to the thirteen cycles of the female body, then one soon falls out of step with the seasons. The year based on visibility periods extends for 384 days. Yet many people have reckoned in this manner, and many people still do. Siberian folk are said to have counted thirteen visible moons to the year. The Dogon of West Africa were so doing at least as recently as 1933, when they were visited by Monserrat Palau Marti. They calculated thirteen months of twenty-eight days with successive weeks of five days (based on the hand and its five fingers). These were apparently reckoned approximately and without reference to the actual phases of the moon—rather a glorified menstrual calendar when all is said and done. The Hopi and Zuñi of the American Southwest reckon by the visibilities, thirteen visibilities to the year. The effort to bring lunar calendar and the seasons into accord is the responsibility of the priests and, according to ethnologist Florence Hawley Ellis, the occasion of no little consternation each year.

The twenty-eight-day-visibility month may have inspired some of the ancient builders of "medicine wheels" in Canada and the northern United States. Reminiscent of the circular observatories erected in Neolithic England for the prediction of lunar eclipses and the sun's position at solstice, the medicine wheels appear to astrophysicist John Eddy to have stellar alignments but may also reflect, he says, the lunar cycles. One of these can be seen in Wyoming, its twenty-eight stone spokes intact.

The thirteen visibility months of the year (and of the female cycle) appear to find resurrection among the Maya, who discarded no temporal calculation but simply added old to new in a complicated network of interlocking systems that continues to boggle the twentieth-century mind. The thirteen-month lunar year found apotheosis in the wheel of thirteen numbered days, which fitted into a larger wheel of twenty named days, which, in turn, meshed with the solar wheel of eighteen months. There were thirteen hours to the Mayan day, and there were thirteen layered heavens above the earth.

We do not know whether the horn held by the Laussel "Venus" commemorates the thirteen lunar visibilities. Alexander Marshak has not discovered among the plaques he has studied the tally of a full year. The evidence, however, does quite clearly indicate accurate understanding of lunar phases, including the dark days. The complete monthly cycle, from first crescent to first crescent, is 29.5 days, and since there is no such thing as a half day in time, one figures the cycle as thirty days in one period, twenty-nine in the next.

A clutch of calendars

If one observes the moon carefully each night and records the visible lunations, phase by phase, the year comes closer to twelve cycles than to thirteen. Yet calculate twelve lunar cycles to the year, and one is still in jar with the seasonal round. Three hundred fifty-four days then comprise the year. Efforts to compose the differences have been as fascinating as they have been complicated. And a little poignant, besides. An insertion of an extra month at stated intervals has been one answer.

Herodotus tells us that the Greeks of his time added a month every third year so as to follow the seasonal round. The more scientific Egyptians, he added, had only to insert five extra days at the end of each year. But they had long ago switched to a solar model, one which would be independently invented by the Mesoamericans.

When the Romans adopted this very sensible scheme, they made improvements which required only one additional day for what we call "leap" year. Their months varied in length, even as do ours. And still there were accumulating errors. The Julian calendar exceeded the true solar year by eleven minutes and some seconds, thus jeopardizing the proper dating of Easter. The Gregorian revision, achieved in 1582, is in use today, and critics still abound.

The religious calendars of the Hebrew and Islamic worlds remain oriented to the moon. The calendar prescribed by Mohammed gains a year on the Gregorian calendar every thirty-two years or so.

A celestial sex change

The recognition that the sun in its seasonal travels—dipping low in the south at winter's beginning—could be used as a more accurate calendar peg may have occurred rather early in time but appears to have been slow to catch on and, as indicated, was never universally accepted. Whenever and wherever the solar cycles encroached on or replaced in importance the lunar calculations, the sexes of the planetary bodies underwent alteration. With a shift of emphasis, the sun became a god, or in the case of the Greeks and Chaldeans, a chariot driven by a god. And the moon, complete with all the earlier attributes, became female. We find a host of lunar ladies, compassionate and tender, in charge of pubescence and motherhood. Yet they are also mistresses of the animals and mighty huntresses. Thus they sprout or wear the curving bovine horns and carry the curving crescent bow. The Greek Artemis—Diana of the Romans—comes most readily to mind.

The transition to Artemis must have come by way of Crete, ancient home of the Mother in her many guises, and before that, surely, there was Çatal Hüyük. In that long-buried city we see the Mother, luxuriantly fleshed, embodying the ancient principle of life and its increase: the life of animals, the life of the crops (one little figurine was found in a grain bin), and human life as well. Throughout Neolithic times and Neolithic space, she would carry this meaning. Her images and symbols are sometimes combined with those of a male deity, more frequently with the curving horns of aurochs and ram, which appear to be his (lunar?) emblem. The lavish Mother and the horned moon twenty thousand years after the Ice Age sculptor had made an identical connection in the cave of Laussel. Can the power of metaphor survive so long, transferable yet unchanged? Why not? The Paleolithic "squat," that ancient attitude of prayer, survived in Çatal Hüyük as a collection of pendants for a necklace.

Five thousand years after the Anatolian florescence, the goddess of Crete had herself assumed the horns. Carved elegantly or crudely, her plumpness constrained by fashion, her arms upraised in blessing, she can be seen in some representations wearing the "Horns of Consecration," a stylized motif which also adorned the palace walls of Knossos. A second Cretan motif, the double ax, or labrys, may also echo in its curving lines the crescent moon, eventually the moon of the Mother, celestial consort to Father Sun, as the earth was everywhere the Mother's body, to be quickened by Father Sky.

The celestial sex change has been reported for Mexico, where among the Maya the moon was female, while the Aztecs, recounting the days of creation in Teotihuacán, spoke of a male moon. To this day a certain ambivalence can be detected among Pueblo people of the Southwest. Opinions about the moon's gender are divided half and half.

A moon for all seasons

American popular usage has over the years cherished a masculine image and associated it, not inappropriately, with lovers and the mating game. *Moon, spoon, June, tune*—they have provided the lyrics of a hundred romantic ballads. Now that we have put a real man on that silvery orb, all the old metaphors, the genders, the deities have tended to dissipate. The moon has become just another piece of high-priced real estate, orbiting, without improved roads or the waterworks that would make life there tenable.

Yet for all its loss of romance, for all its demotion as a ready reckoner of time, the moon still commands the night and the rhythm of the ocean tides. And while it does not bring to women the flow and the ebb of bodily tides that announce the quickening of new life, it still bears those ancient connotations. Its cycling phases recall a time when changes of the female body—blooming pubescence, fertile maturity, and barren age—were the only recognized cycles of life, mysterious in their causes, predictable in duration and in effect.

"To everything there is a season," said the Preacher in Ecclesiastes,

> and a time to every purpose under the heaven: A time to be born, and a time to die; a time to plant, and a time to pluck up that which is planted . . .

For all the cleverness of the human mind, however far that mind can be projected, whatever self-transcendence can be achieved, the human seasons still remain. Time and tides wait for no man and will bear us all home. Home to Mother. Home to Father. Or to Whom It May Concern.

PART THREE

BODY RITE

9

DENYING THE FLESH

In Command of the Appetites

Jonathan Swift in one of his satires referred to man as "the animal who eats when he is not hungry and drinks without the provocation of thirst." And, as others have added, "whose amorous passions know no season."

But also, one must remember, man is the only animal who declines to eat when hungry and who can willingly forgo the pleasures of the flesh. The only animal capable of renunciation, of starving the body to feed the soul.

How humankind arrived at the notion that sexual sobriety was more admirable than sexual excess, that a rumbling belly in time of plenty was more to be honored than the same condition when commons were short, is certainly anybody's guess. The scientific school of hard knocks would doubtless envision such rules as the natural sequel to a calamitous period of primitive hedonism. The bulk of the ethnographic evidence, however, suggests a mute appeal to the Powers made through the human body. *You give, I give up.* This is the message of the fast.

Through all the elaborations of custom, through all the changing lineaments of the Powers, always there has remained this belief: that giving up is both an approach to holiness and holiness itself. Renunciation is as standard an accompaniment to the religious act as cleanliness. Often it provides the only marker of the holy in events and situations that might seem at first blush not only thoroughly secular in tone but even a trifle seamy.

The ascetic capitalist

Max Weber, the much-admired German sociologist of the early 1900s, saw in the capitalist economic system not the vulgar materialism of which in his day (as in our own) it often stood accused, but the lingering asceticism which spoke of religious underpinnings. In a long essay titled *The Protestant Ethic and the Spirit of Capitalism,* he traced our modern system back to its Calvinist beginnings. For members of this early Protestant sect (many of whom came to our shores as Pilgrims), hard labor in a rational calling was God's moral imperative. And if one prospered thereby, why, such was the mark of divine grace—provided one did not squander prosperity in idleness or foolish pleasure. One might not eat overmuch or sleep overmuch, and certainly one might not dally with the opposite sex. It could well be argued, of course, that every new sect is born in a spasm of righteous puritanism out of orthodoxy, worldly and lax. And this is as true of the political sect as it is of the more obviously devotional one. Surely the gray and moralistic China wrought by the Cultural Revolution pursued the goal of clean living and self-denial as fervently as did Puritan New England in its time.

Still, holding Weber's thesis in mind, it is instructive to find in a primitive setting echoes of the Calvinist belief in unrelenting effort and abstinence and in wealth as a sign of grace. It is an echo noted nearly thirty years ago by the anthropologist Walter Goldschmidt.

Long before American capitalism arrived in northern California, each Yurok prayed nightly amid gushing tears, "Let me be wealthy! Let me be wealthy!" Then as now, however, God helped those who helped themselves, and the Yurok hunter strove through industry and artful trading to lay up treasure in his carved wooden boxes, if not in Heaven. But industry was never enough. He must learn and practice the airs and graces of a rich man. He must cultivate slenderness by exercising the rich man's command of his appetites. And this was tested daily in the sweat house, the communal fires of which he stoked with hard labor. The sweat house exit was deliberately made a bit too small. The man who had eaten overmuch at his latest meal could not contrive to wriggle through.

If in the Yurok belief dentalium shells and women did not mix, neither did women and deer, those animals so prized as food and for their skins. The successful hunter was the abstinent hunter. And if his lack of skill or luck, or both, made him a poor man, his peers would surely point to the over-many children frolicking about his hut and cluck with disapproval at his obvious lack of self-control. "Like a dog," they would mutter balefully. "Just like a dog."

The ascetic hunter

Throughout the world of the hunter, abstinence is a prime directive. For the animals who willingly give themselves to feed humankind are prudish when it comes to human sexuality and must not be offended. What is natural among themselves—coitus without the dread of incest, birth without the horror of blood—they find repugnant in humankind, among whom a proper decorum, a proper order, should prevail. The careful man, therefore, sleeps apart from his wife the night before a hunt is planned. Bushmen of the African Kalahari are even reluctant to speak of sex in front of women lest the hunter's abilities be impaired. Very old men can explain these things to the inquisitive anthropologist. Younger men, never.

Among some folk such as the Washo, a Great Basin people who once lived near Lake Tahoe, prohibitions were extended to the ceremonial harvest of piñon nuts, their staple food for the winter months. The present-day Maya farmer, who hunts seldom if at all, transfers the hunter's care for prey to the trees that must be removed to make his milpa—his cultivated plot. There must never be too many cut, never wanton destruction on a field larger than can reasonably be hoed. And before planting, the farmer keeps vigil before an altar bearing a bowl of sacred corn. All night he is there, forswearing the comforts of the connubial couch. Indeed, he may remain abstinent for many days thereafter until the seeds have germinated and are rooted at last.

Among hunting people everywhere, the wife must refrain from touching her husband's hunting or fishing gear. She must take care that he is not afflicted by the taint of birth or menses. Should she be careless or, resenting the inconvenience, skimp the proper observances, she invites serious illness. Worse yet, she may jeopardize hunting for everyone in her community. In her personal lapse there lurks public catastrophe. Here is the stuff of sin and retribution and, for all the modern theologian's label of superstitious nonsense, high tragedy as well. Only confession can provide a remedy.

Knud Rasmussen, the Danish explorer and anthropologist, himself half Eskimo, witnessed many years ago such a confession on Baffin Island. A woman, desperately ill, was visited by the shaman and a crowd of her fellow villagers. As the shaman conducted the inquisition, they chorused in their turn, "It is a little thing. Let her be released from this burden, for this offense. Let her be well."

Relentlessly the shaman continued to draw from her, one by one, a litany of sins, some twenty-five in all. Twice she had combed her hair when that was forbidden. Twice she had touched the things of the dead. All the rest had to do with sex and food. She had hidden the fact of a miscarriage, tainting her husband thereby. She had lain with men when forbidden to them, touched their hunting gear, altering it into garments for her personal use. She had handled the things of Nature—bone, fur, leather, birds—when she must not touch these things. She had eaten foods out of season, the foods forbidden twice over.

"Even for so hardened a conscience," intoned the shaman, "there is release. . . . Return to life." And the assembled throng departed, said Rasmussen, secure in the belief that the woman had been healed and would soon be well. Sin and redemption.

Chastity and concentration

It was not so much the dread of sin as a bent for conservation that elevated chastity among the Cheyenne to a supreme virtue. In the view of these North American migrants to the Great Plains, all the world's energies, including the sexual ones, were limited. Easily depleted, difficult to restore. They should therefore be husbanded carefully, never squandered. And so it was in the old days that the young man of good family might offer a vow of celibacy on the birth of his first son. He would not engender another for seven years or fourteen; yet all the while he slept beside his wife. Certainly that lady would remain untouched. Cheyenne women wore chastity belts but were guarded even more securely by their ironclad reputations. For the prescribed seven or fourteen years, the sexual energies of both parents would be channeled into their child. Anthropologist E. A. Hoebel reports that such vows did indeed appear to produce offspring who were outstanding, even among the upright Cheyenne, for their superior intelligence and character.

One encounters everywhere the belief that sexual activity is not only incompatible with great doings and great events but is also a distraction, siphoning off force and energy needed elsewhere. Even the easygoing Polynesians suspended dalliance during the seasons of ritual and enjoined chastity on certain craftsmen at their creative labors.

The makers of religious masks were everywhere removed from the temptations and distractions of life at home lest divine inspiration be denied. The Dogon called mask- and costume-making "the menstruation of men." By thus rendering the artist unclean, they ensured his social separation.

This is not to deny the practice of ritual license or the belief, held in many settings, of the sacred efficacy of the generative act, properly ordered and constrained. Herodotus insisted that the Egyptians were first to banish sexuality from the temple and that all other people (excepting, of course, his own Greeks) allowed the two passions, spiritual and carnal, to coexist without let or hindrance. The notoriety of temple prostitution as practiced in Babylon had doubtless biased his usual objectivity. A brief review of the record, both historical and ethnographic, gives the palm to abstinence.

From initiation rites to vision quests, from the offering of first fruits to animal sacrifices, all have required congregational chastity as a condition of grace. So have war, which among some folk is another kind of rite, and games, which are another kind of war. Tribal folk from Africa to the New World went into battle with strength intact, safeguarded from female wiles and sapping female allure. Or, at the very least, their commanders were so defended.

Chastity old and new

The South African Zulu, once mighty warriors, have transferred to the purposes of Western soccer their pre-battle rites and seclusion. On the night before a game, reports ethnographer Norman A. Scotch, teammates camp together, naked, around a fire, avoiding thereby the lure of women and alcohol, the danger of enemy players who might weaken a teammate with sorcery or infect him with ill luck. During the night the players are treated by the team diviner, who makes ritual incisions on all their joints and directs his own powers of sorcery against the enemy team. Next day, the players travel together to the game, entering the stadium in tight formation, together to do battle.

This is single-purpose celibacy: *You give, I give up.* The religions of some great states have typically set apart special individuals or groups whose all-purpose, lifelong renunciation served to raise the level of public sanctity and guarantee general holiness. The Roman Vestal Virgins, the Inca Virgins of the Sun, the Virgin Princess throughout Polynesia, the Priest-Chief of the Dogon offer examples. Aztec monks and nuns, Aztec seminaries and convents, startled and perplexed conquistadores by reminding them of similar institutions in Catholic Spain. Startling, too, was Huitzilopochtli, chief of the Aztec gods. While not the product of virgin birth, his was definitely a miraculous conception.

Monks and nuns are also to be found in those parts of the world where Buddhism reigns. And Hindu custom—more a way of life than a religion—which has no place for the bachelor or spinster, heaps honors on the person who renounces sexuality to the greater glory of God. A young bride and bridegroom are traditionally revered as incarnations of the god Siva and his wife, provided they can remain chaste for three nights at least. Six nights are better, some observers declare. Better still are twelve, and a year best of all. For a Hindu male the ideal life is strictly compartmentalized. First as student, obedient to one's teachers; next as householder, husband, father; next as a married recluse; last as the ascetic, all worldly claims abandoned, without home, without wife, with naught between life and death but the unending contemplation of God.

Although Saint Jerome seemed to locate marriage uncomfortably close to sin, religious chastity has never attracted wholesale conversion in the West. Today even the inadvertently continent attract a raised eyebrow or two. Current wisdom tends to regard the slightest bridling of the appetites as a hazard to health and perhaps sanity as well. Still, sexual asceticism in pursuit of purity and success (which may be another sort of sanctity) blooms unrecognized in some very unlikely temples.

Visit the boxer training for a title bout and you will find his camp a veritable monastery. Visit the professional football team in training and you will find, says William Arens, that the men are sequestered during the week. When the season begins, just the reverse is true. The men are sequestered on the weekends. They travel to the game on the same bus, girls strictly forbidden, and they enter the stadium together. Should wives tag along to the out-of-town games, they might well find themselves evicted from their hotel rooms on Saturday night. As one player told Dr. Arens, "Tonight we're putting the girls in one room, and Danny and I are sharing one. It's better for the girls to be away from us tonight. We're always grumpy and grouchy before a game." And also, it would seem, ceremonially abstinent. *You give, I give up.*

Starve the body; feed the soul

Denial of the sexual urge may be felt less keenly in those parts of the world where its ritual appearance is invariably accompanied by the denial of sustenance. No fuel, no performance.

In terms of sustenance, few people have been so poor, so poised on survival's very edge, as to eat everything edible. Be it ever so humble, a tribe will certainly identify forbidden fruit. For while food nourishes the body, its denial feeds the soul, setting off culture from nature, human from animal.

Not all animals, it would appear. Among Australian Aborigines, one's maternal clan is identified by what its members do not eat. This is the animal, or other edible entity, which is one flesh with them, which is their totem. Thus the shared meal does less to cement family ties than does the shared taboo. The family that abstains together remains together.

Among some Polynesian peoples to whom totemic beliefs extend, a husband (necessarily of another kin group) may not eat of his wife's totem lest their conjugal relations prove incestuous. Food, after all, makes semen. So watch the diet.

If denial of food defines family, it also reinforces family lessons to the young. Among many hunting people, choice cuts of meat, the tastiest fish, may be denied young boys so as to encourage them to develop hunting skills and an accelerated entry into man's estate. Yet the first large game bagged by the young hunter—the kill which gives him adult status—may not provide his own dinner. It is for everyone else to eat. So, too, for the young girl at the first gathering after her puberty seclusion. Thus do they learn, man and woman, the social virtues of sharing.

The denial of food brings blessing and brings luck. It is a wordless prayer to the Powers. "Do for me as I do without. See how pure I am, how free of bodily necessity. See how well I obey your laws. Reward me therefore."

Do without when?

The laws of denial are many, some meant for small groups and individuals, some for the social order, however that is reckoned. The general laws invariably prescribe the times for avoidance or for fast, the circumstance of fasting, and the place. Dietary proscriptions of the Eskimo are rigidly seasonal. One may not, for example, hunt the caribou when marine animals are in the ascendant. One may not eat forbidden meat out of season, however thoroughly frozen, or even gnaw a bone thereof. Nor may one handle fur or skins of the out-of-season animal, for this offends the Lady of the Sea, and she will bring storms and bad luck and a dearth of game. Yet there are private lapses from this stern code, lapses which afflict sinners, as Rasmussen described, and for which penance is decreed.

Such separations are known to many hunting folk. The

Yurok never mingled whale meat with venison on a single platter or included bear meat and salmon in the same meal. And when hunting one of a pair of opposites, a man always took care to disparage the other, loudly referring to whale meat as "rotten wood" if he thought a deer might be within earshot. Such separations were retained by agricultural people, who avoided a ripe harvest or growing plants until they were made ritually safe for consumption. In the Book of Leviticus there is one rule which states:

> Thou shalt not sow thy field with a mingled seed: neither shall a garment mingled of linen and woollen come upon thee.

Hebrew dietary laws, observed today by the Orthodox, prohibit the serving of meat and milk at the same meal. The *Jewish Encyclopedia* suggests this may be older than rabbinical interpretation of the law, probably deriving from Temple practice which avoided mixing dishes which required different ceremonial treatment.

The grand year rituals, whether of celebration or penitence, have historically included fasts. The Creek "Busk," a harvest festival–New Year's Eve–born-again Baptism all rolled into one, is really a corruption of the Muskogee word meaning "to fast." And everyone did, for days. Through it all, men slept in the town square, apart from their wives. Aztecs celebrated their holy days fasting, and fasted in fear and dread during the five intercalary days, the evil season of the Aztec year.

Dare we bracket the elaborate asceticism of the great world religions with that accompanying a Creek Busk or an Aztec sacrifice? Why not? They spring from the same source: the human desire to acquire merit and grace, to be pleasing in the sight of the Powers (however these are named and numbered).

The devout Moslem concentrates his fasting in the lunar month of Ramadan, the ninth of the Islamic year. Though probably of very ancient sanctity, Ramadan is believed also to be the month during which the Koran was dictated to Mohammed.

Catholics of both Roman and Eastern persuasions reserve extended fasting for the Lenten season preceding Easter. Meat is the primary omission, as it was until recently every Friday and sometimes Wednesday as well. During the last week of Lent the Greek Orthodox Church additionally forbids all dairy products. One contents oneself with bread (made without milk) and vegetables. On the day before Easter, one eats nothing at all. And, as old congregants insist, the lights, the colors, the incense at the midnight Mass of Resurrection are apt to be not a little hypnotic in effect.

So are the seasons of holiness defined and separated from the dailiness of life.

In the matter of denial, no one could outdo your everyday, garden-variety Hindu. And of the Hindus none was more zealous than the man of highest caste, the Brahmin. (In isolated villages, this may still be true.) He was expected to fast every day until noon, even when working hard. The days of new and full moon required additional fasting. So did the tenth, eleventh, and twelfth days of the lunar month. On the eleventh the fast was total. On the thirteenth, an especially unlucky day, one could not break fast until after sundown. On the fourteenth day of the moon equivalent to our February, no one might eat, drink, or sleep for twenty-four hours. And for all the great rituals, more fasting was enjoined, along with perfect celibacy.

Do without whom?

As important as fasting was to the pious Hindu, more important still was the company one kept at meals—one's companions, those with whom one broke bread. On pain of defilement those must not be of a caste lower than one's own. There were rules stipulating from whom one could or could not accept water, who could or could not cook one's food. Breaking of caste barriers (except by holy men whose utter asceticism placed them outside the rules of separation) was polluting, not only to oneself but to the community at large.

Foreigners of every sort were kept at arm's length by the fussy Egyptians, who would not, Herodotus reports, kiss Greeks on the mouth because of their unfortunate dietary preferences, notably their fondness for beef. Whether Greeks were banished to separate dining facilities we are not told. Certainly the Hebrews, even those of high rank, were not considered proper dinnertime companions. (And vice versa, one might add.) Of Joseph's reunion with his brothers, we find this account in Genesis 43:32:

> And they set on [bread] for him by himself, and for them by themselves, and for the Egyptians, which did not eat with him, by themselves: because the Egyptians might not eat bread with the Hebrews; for that is an abomination unto the Egyptians.

In bed and board are always drawn the ultimate lines of exclusion. During the civil rights struggles of twenty years ago, it was the desegregation of schools and restaurants that provoked among whites the loudest protests. The one threatened a propinquity which might lead to miscegenation; the other conferred that ultimate equality, the breaking of bread.

Throughout Oceania, and especially in Polynesia, taboos of companionship focused on board and never mind the bed. Hanky-panky in the bush was in the normal order of things. But let a man and a maid eat together, and scandal ensued. In Hawaii considerably more: the penalty was death.

Women might not eat with men, nor might they eat the foods of men—pork, fowl, turtle, several kinds of fish, coconut—for these were offered in sacrifice to the gods of men and were eaten by men in meals which were sacramental every time. It is true that women, some of them, were born into high rank and exercised great power. They also had their own goddesses. Nevertheless, a proper division of diet and rite was the cornerstone of the Hawaiian religion. When in 1819 the great Kamehameha was succeeded by a young boy, the dominant women of his immediate circle persuaded him to eat with them publicly. And when Mount Pele failed to erupt and the grim idols remained securely on their altars, the religion itself fell into disarray and was never again reconstituted.

Do without what?

Nearly every people recognizes certain foods as fit for consumption and others not. The choice usually has little to do with nutrition. "One man's meat is another man's poison" is the operative proverb in this instance. We gag at the prospect of perfectly edible worms, lizards, and meat of dog or horse. Frog legs and snails, those ultra-French delicacies, are for us tastes acquired with effort. Egyptians found beans revolting; so we are told by Herodotus. It all depends. And often the requirements of ritual determine when the gorge will rise and when it will not.

The prohibitions of Leviticus center on animals which do not chew the cud and/or lack a cloven hoof. Dogs, swine, and camels are thereby rendered unclean. So are hyrax, rabbits, and the pawed animals such as weasel, mouse (several mentions of this occur), ferret, mole, and the various reptiles. Birds of prey are excluded, fish without fins and scales are excluded, but with locusts, beetles, and grasshoppers anything goes.

Among North American Indians, the Creeks had a long list of forbidden foods. Birds of night and of prey; flies, mosquitoes, gnats; animals which themselves ate "nasty" food; all amphibians and most reptiles; all carnivores except the bear. One might not eat animals newly weaned, and must "bleed" one's prey and cook it very thoroughly thereafter. To this list the Iroquois added the flesh of pregnant animals and those which frequented burial grounds.

"There is no accounting for taste"; so the old saying goes.

"Oh, yes, there is," insist students of humankind, "and there are several avenues of explanation." One can, they say, use the historical record if this is available or can be reconstructed. There is the way of meaning, if this is known or can be reconstructed. What, for example, are the reasons given by the people in question for excluding certain foods? How does this mesh with the whole fabric of belief and of reality?

This last is an approach labeled too "mentalist" by the proponents of ecology and of function. Any custom bearing on the breadbasket must, in their view, have some clearly adaptive value. It must serve to guarantee the dietary resources of the group, or its health and safety, or at the very least it must in some way work to hold the group together.

The Jivaro prohibition of deer and tapir (demons, say the Jivaro, sorcerers reincarnate) has been called by the ecologists a response to scarcity. Conservation of an endangered species, so to speak. And it is true that the large animals are not only rare in appearance in that part of South America but also wary and difficult to kill. The tribe which settles permanently in one location without troubling to forage afar will have to make do with small game and rodents attracted to their cultivated fields. What is not available, as one authority puts it, soon becomes unfit to eat. Even when it returns in some abundance, it thus stands magically protected from the human predator. (The Jivaro and other people of the Ecuadorian and Peruvian Montaña prohibit vultures—"They eat nasty things"—but little is made by specialists of this taboo.)

Explaining in ecological terms the religious prohibitions of committed agriculturalists is somewhat more intriguing and of endless interest to the experts. The best known of these taboos involve the cow in India, sacred and never to be slaughtered, and the pig, forbidden to Jew and Moslem alike. Anthropologist Marvin Harris, writing of the Indian cow, declares that it is of far greater utility in traction, as a scavenger, and as a producer of dung (at once fuel, fertilizer, and mortar) than as food. As for pigs, he says, they provide nothing but meat—no hair, no milk, no usable dung. Since they do not sweat well, they need a moist and shaded habitat. And when, during Neolithic times, Middle Eastern forests were everywhere replaced by farms, pigs lost both shade and forage. To be raised to eating size, they then had to be nourished with essentially human provender and became expensive luxuries, competitors in terms of food.

Yet any pig breeder can tell you that these animals eat anything, including human waste. (Indeed, they have often been described as "walking sewers.") That they have been proscribed at least in part because of filthy eating habits, Dr. Harris discounts as a reason. Melanesian pig fanciers are not, after all, put off by such uncleanliness. (But one remembers the Creeks, who considered that an animal's diet figured in rendering its own flesh edible or not.) That pigs carry disease Dr. Harris discounts also. So, he says, does the underdone meat of many another animal.

Predictably, his explanatory models have been challenged even by other ecologists. No single answer, it seems, suits everyone. Yet specialists are equally averse to the holistic ones. Only a genius, some psychologist once said, can encompass at one time more than two fence posts in the mind's eye. That apparently applies to causes as well.

In all the customs and taboos of the ancient Middle East (and perhaps even farther afield), it is hard to ignore the powerful influence of Egypt. Herodotus, who visited that land in the fifth century, wrote at length about Egyptian cows (all of them sacred to Mother Isis), which might not be killed under penalty of law. Bulls and bull calves might be sacrificed by the priests, but only those animals utterly without blemish, possessing not so much as one black hair. Remembering Egyptian conservatism, we may suppose this to be a taboo of long standing. Swine, he wrote, were considered so unclean that the mere touch of one would send a man running pell-mell to the river to dunk himself, clothes and all. Keepers of swine were men apart, barred from any temple or from any maid save the daughter of another swineherd. Yet Egyptian priests did eat this forbidden flesh as sacrifice when the moon was full.

> They give a reason [said Herodotus] for offering pigs at this feast and abhorring them at other sacrifices, but it is not seemly for me to tell what it is though I know it.

So much for the confirmations of history.

Doing without, the special cases

If tracing the origins of general taboos reveals only a twist of possible answers, explanations of the individual cases are even more obscure. The personal prescriptions and proscriptions dictated in dreams; the songs, the magic formulas, the compulsive acts which must be performed; the foods which must be forsworn—must surely be as many and as various as there were people who based their lives on such rules. And there are as well the cultural regulations prescribed in special circumstances. During illness ("Feed a cold and starve a fever" is one of our own), in time of mourning, puberty, parturition, war.

Thomas W. Neumann, concentrating in the main on aboriginal North America, noted a common factor: the salt taboo. And he concludes a logical advantage, whether clearly recognized or no. In times of physical or emotional stress, he points out, the body tends to retain its salt. This may lead in turn to a retention of body fluid and a hypertensive condition.

Now salt is a valued, an indispensable commodity only among horticultural folk less dependent on meat and fish than are hunters and gatherers. And so, he says, the standard taboo for the hunter under stress is invaribly fresh meat and fish, if not salt outright. The possible benefits of the salt-free diet at menstruation are easy to see, and for pregnancy as well—though it is certainly true that not every expectant mother suffers from edema and hypertension.

But fathers, too? Often the taboos imposed on Mother are shared by her husband, as are the rigors of confinement. In some cultures, indeed, Father takes to his bed immediately following the blessed event, moaning and groaning with a prolonged metaphorical afterbirth while Mother tends him and Baby as well. Among the Washo, food taboos came then, not during pregnancy. Everywhere the rigid round of avoidances and ritual focus on the safeguarding of the child for whom both parents are responsible, actually and spiritually.

It is hard to see the advantage of a salt and meat taboo (often including water as well) for the sick person. The lack of adequate nourishment is apt to hasten exit rather than recovery. One suspects another reason for denying the sick, the pregnant, the girl in her moons. They are alike offensive to the game animals, whose flesh should not be defiled. The same connection may well have been applied to those in mourning, those touched by death. The Ojibwa widow was thought capable of blasting the earth with her very presence and was denied fresh meat or fish.

Elsewhere the simple omission of salt for the grief-stricken appears as a lesser denial, something not quite so rigorous as the fast but not yet pleasure, either. The lesser denial is echoed in happier circumstances. The Hindu bride and groom, impersonating Siva and his wife, are for three days not only chaste but saltless. A true fast, enjoined in all other circumstances, would not be appropriate here lest the couple lose strength for the eventual connubium. "Eat," is the message, "but don't enjoy the food."

The prolonged deprivation of salt can trigger severe muscle spasms, can impair judgment and cause eventual derangement. Dr. Neumann thinks these states may have been expected and desired during puberty when initiates, like those in the tribal Southeast, were confined for long periods. (But parents, too? The Ndembu religiously avoid salt while the circumcision wounds of their sons are healing.) It must be remembered, of course, that Creek and Cherokee novices were fed intoxicants, which might also account for the sort of lunacy described in old Colonial accounts.

A state of fierce derangement may be desired in battle. In North America, warriors regularly fought while fasting. Doubtless the custom helped them to develop the sort of last-ditch, second-wind capability that made them such formidable opponents. Yet Dr. Neumann himself wonders how strategically effective were the long-term effects of prolonged fasting, broken only by a little parched corn to stave off collapse. No meat. No salt. Except for the fasting prescribed for the Japanese samurai, we see little of such rigors in military circumstances elsewhere.

Yet fasting arms the warrior with spiritual weapons. Such at least was the belief of the Creeks. On expedition, the warrior took no food for three days and three nights. He did not rest his limbs, did not sit, could not even lean against a tree until after sunset. The more strictly he adhered to this stern code, the more virtue he could claim. And the more virtuous he was, the greater would be his ultimate success. If war, then holy war.

Virtue is power

In the final analysis, the body is the last and perhaps the only command post of the soul. Unable to manipulate destiny or fellow beings, we can at least direct the destiny of our own flesh. Victims of anorexia nervosa, with their horror of becoming fat, seem to find transcendence in slow starvation. There is power in denial, a fact which has not escaped the so-called primitive peoples of the world.

Consider the Fox, or, as they called themselves, the Mesquakie, "Red Earth People." Originally they lived in eastern Wisconsin but were driven out first by other Algonquian-speaking tribes and later by the French. In the late 1800s the anthropologist William Jones, himself part Fox, wrote about his people and their uses of the fast.

If a young man had ambitions for leadership, he notified the Great Spirit by fasting. If he wished to be great, he learned how by fasting. If he wished to be a mighty hunter, he achieved his desire by fasting. He fasted for visions, for revenge, for war. He fasted in mourning and in penitence. The cautionary tales of the Fox often begin in this way:

> Two youths blackened their faces and fasted during the course of one whole winter and not until spring did they cease from fasting in black.

It was considered proper to construct a "fasting instrument"—a length of peeled, dried basswood which was burned bit by bit to provide the charcoal for blackening. One never fasted or told stories with the face in its natural shade of skin. When the stick was all used up, the fast was at an end, and presumably the celibacy as well. For the fasting youth was constantly referred to as "clean," as one who "did not bother with women."

The Fox thought it unwise to fast beyond the granting of a boon. Such diligence in denial gave one a charge of power impossible to contain—power that might kill others and harm oneself. Power. That is what it was all about. Strength through holiness.

This is power known well in India, whose people have come to honor and obey the ascetic far more than the man of action or even the man of knowledge. To those who command the cravings of the flesh, who fear not the little deaths of renunciation, are ascribed courage in all else. Certainly this was true of Mohandas Gandhi. Given his admiration for the Jain sect, which decrees reverence for all life save one's own, he might well have ended his life in slow starvation had it not been cut short by an assassin's bullet.

People of the West have not been notably enthusiastic about voluntary denials of the flesh. Except, of course, during that spasm of holiness that marked the Middle Ages. But little by little, the customs of the East have gained adherents in this hemisphere, beginning doubtless with the British rule in India. The first Vegetarian Society was founded in England in 1847 and was to number among its membership the young Mohandas Gandhi during his university years, as well as the irrepressible George Bernard Shaw, who in his later life would scandalize his circle by taking liver shots.

Asceticism triumphant?

During the sixties, countless American "flower children," refugees from the drug cults, embraced Eastern mysticism together with the religious diets prescribed. Many other young people, unmoved by ideology, were nevertheless persuaded to alter their eating habits, to forswear meat, to do without. It is now a veritable movement. One has only to note the number of college catalogues which announce the vegetarian option regularly offered in their dining halls. These meals are prepared with Brahminical attention to detail. For the vegetarian young reject not only meat but anything cooked in animal fat or juice. Even a frying pan improperly scoured of same is taboo. It is ritual defilement without the ritual.

"Taboos," said K. L. Brown of the University of Manchester, "aren't simply ancient but continually get created." And here is an example. Perhaps some ecologist of the future will link it to the high price of meat. Vegetarians themselves, those without the religious rationale, are apt to cite moral reasons or scientific ones. Meat is poison to the system (and maybe milk and eggs as well). Meat is not in the natural order of things: primates are vegetarians, and we are primates. Eating meat is a violation of rights, animal rights.

The puzzled skeptic is often moved to point out that, while some meats increase the body's cholesterol level, not all meats do. He also wonders about baboons, who relish whatever small animals they encounter, and about carnivorous chimpanzees. Not to mention the fact that our own species seems to have got its start via the hunting way of life. And if animals have rights, why not plants as well? Luther Burbank, the horticultural genius, secretly believed that plants had a sort of vegetal sentience and reacted violently to the prospect of their own impending demise. So what's left to eat?

To chop logic, however, is to miss the point of doing without, which is, after all, purity and power. On this world of giants—giant business, labor, government—individuals have little impact. They are hard put to stand out in the crowd, let alone to make their mark. But they have command of their bodies and the power to say no. Denial places them in that special company of the elect. They are pure, righteous, holier than thou. Never mind the sanctions of a deity. Like the atheistic asceticism of the early Jains and Buddhists, doing without carries its own rewards.

It is self-abnegation in search of a soul.

10

MORTIFYING THE FLESH

The Sanctity of Pain

Over the Malaysian jungle, thunder pealed and rattled in a frightful barrage. Lightning flared again and again, brilliant ropes linking earth and sky. Below the trees a band of Semang gatherers huddled fearfully in their flimsy sheds. Suddenly a woman holding a bamboo splinter rushed out into the rain. Gashing her shins with this improvised knife, she collected blood and rainwater in a bark container, poured some on the earth, and then threw the rest heavenward. "You there, Grandmother," she shrieked, "climb up and make known to your grandsons that I am paying the debt!"

Which debt? The ethnologist Father Paul Schebesta, who traveled with the Semang during the early 1900s, never learned. The rules which then governed Semang life—and may still do—were many. Mostly they dealt with sexual offenses and taboos, with food taboos, the treatment of forest animals, and with forest etiquette. Perhaps one had become overenthusiastic in conversation or greeted a husband, long presumed tiger-rent, with joy too loudly expressed. For even this careless infraction Karei, the Thunderer, might inflict punishment. And to avert his wrath, women offered payment in pain. Better a little now than a lot later: this seems to be the meaning of the slashed shins. It is a price that only humankind can pay. Animals may be magical. But no animal can choose to pay. No animal can invite pain and use it to control the natural, the powerful, and the feared.

Certainly no animal uses pain in coercion. The Amahuaca do. In their eastern Peruvian hunting grounds, so Robert Carneiro tells us, game is often scarce and always wary. To bring good fortune, better vision, better aim, the hunter may burn his arms with caustic sap. He may hold in his hand a nest of angry wasps until he can endure their stings no longer and must run madly through the forest, still holding the nest. He may rub into self-inflicted burns or wounds the poison of a small frog, suffering for three days thereafter both intense pain and frightful hallucinations. He emerges from these trials a happy man, confirmed in luck and skill.

Rather pain than death

The belief that bargains can be struck with fate, that sorrow and disaster can be averted if one is willing to pay the asking price, runs deeply in the human consciousness. Which of us, standing by the bedside of spouse or child struck ill and perhaps dying, has not bargained silently with no one in particular, "Take me instead. Take *my* health, my strength, my luck. Take anything, but leave to me that which is more precious than life."

We say that; we do not act it out. We do not, like the Arapaho chief, fearful lest his brother die, announce before the assembled camp:

> In order that my brother may get well, let it be known to all spirits that, early in the morning, I shall cut seven pieces from my skin and, in lieu of my brother, I shall bury them.

We do not pledge the equivalent of a Sun Dance during which we expect to have our muscles skewered so as to drag burdens of buffalo skulls. We do not, like Plains women of other times, lop off fingers to procure the safety of warrior husbands. We do not, like the Iroquois dreaming of torment by enemy captors, have ourselves tortured by friends instead, thus forestalling fate. We do not, as in many a Latin American village to this day, vow barefoot pilgrimage to a sacred shrine or crawl to the local cathedral on our knees. We do not bind ourselves to silence with a silver mouth lock or swear to hold a hand clenched until the nails grow through the flesh—austerities of thanksgiving observed on the Indian subcontinent.

Were we to do so, we should be carted off forthwith to the nearest psychiatric ward, where our "masochistic" compulsions could be properly treated and restrained. "You can think but mustn't tell. Above all, mustn't do." Such is the requirement of our culture, stolid as it is, temperate, oriented to the "normal"—however that is currently interpreted. Nothing too much (except perhaps work) is the goal. Neither too much sorrow nor too much joy. At the loss of a loved one we must grieve unobtrusively, at least in public, and return to the mainstream of life as soon as decently possible. Steady on. Stiff upper lip. And all that.

Pain and bereavement

People in cultural settings different from our own have also valued a seemly control of grief. Others have been too fearful of the loved one's shade, too concerned that it take survivors along for company, to give much thought to control. Still others have encouraged, and still do, a passionate display of grief, frenzied and unrestrained. The rage which often seizes those who mourn (and which in our culture is rigidly suppressed) is here given free rein. The ethnographic record abounds with funeral battles—some sham, some real enough to give additional cause to mourn.

Australian Aboriginal custom allows close male kin to fall on the dying man, smothering him in embraces, body to body, head to head, a custom which (as Dr. Elkin remarks) doubtless hastens his departure from this world. The wife and certain relatives (just which depends on the local system of descent and filiation) then gash and hack themselves with clubs and stones. Men may gash their legs, tying the wounds open with hair string to ensure ostentatious scars. Women always choose the head, gashing scalp and cheeks until they bleed freely, turning rage and anguish against themselves. Men are not compelled so to lacerate themselves on a woman's death or on that of a young child. No lesser grief is demonstrated but rather a lesser emphasis on the ritual of grief.

Very little girls among the Dugum Dani, those three to six years of age, are chosen to suffer the pains of bereavement. For the death in combat of a close male relative, little girls lose their fingers. "Why not?" say the Dani. "Everybody makes funeral offerings. Some bring pigs. Some bring shell ornaments. Little girls bring fingers." Not willingly, Dr. Heider reports, in spite of protestations to the contrary. (He heard the wails.) Grown men occasionally slice the tops of their ears or chop off fingers but do so only in an excess of very real grief—often for a wife.

But mourning obligations, those at least that require mortification of the flesh, seem everywhere to fall most heavily on widows. Cheyenne women, says Dr. Hoebel, slashed themselves all over, leaving on their bodies the dried, caked blood until warriors were moved to pity, moved to seek revenge. It may be that the contamination and exclusion which the new-made widow saw ahead lent additional fervor to her grief.

The exclusion of the bereaved—particularly the bereaved female—was customary in America fifty years ago. The widower with a family to support (however devastating his grief) had soon after the funeral to be back to the workaday world with only a black arm band to announce his change of status. And if there were small children, he was expected (whether or not this suited inclination) to find for them a new mother as soon as possible. The widow, however, especially the widow of social standing, undertook a whole list of restrictions—most of them spelled out in the etiquette books of the day—and all represented by variations of the "widow's weeds." For the first months, unrelieved black. Then the addition of jet beads and ornaments. Then touches of white. Then costumes of gray or lavender. Then the normal wardrobe resumed. The whole cycle required about a year to complete. In some Mediterranean countries, until quite recently, mourning was without end. At any gathering the widow stood out like a raven amid doves.

Widow's weeds on the Melanesian island of New Britain were less emblem than mortification. Yards and yards of heavy vine were wrapped so tightly about the legs as to make walking nearly impossible. Of course there was no place to walk. The widow and sister of the deceased were walled up for several months in a sort of closet and allowed out only twice a day to answer the calls of nature. Even when released, the widow was for two years forbidden to garden. Doubtless this was intended rather more as a protection of the plants than of herself. Dread of this sort of contamination and even (as described earlier) punishment might well have caused many a Hindu widow to consider death on the husband's funeral pyre a preferred alternative and certainly a more honorable one. So widespread was this particular custom that it was banned by law of the British Raj.

Pain and politics

As Richard II says in Shakespeare's play,

> For God's sake, let us sit upon the ground
> and tell sad stories of the death of kings . . .

And then, with death's mockery of pomp in mind,

> For you have but mistook me all this while:
> I live with bread like you, feel want,
> Taste grief, need friends. Subjected thus,
> How can you say to me I am a king?

Kings are indeed mortal, but their passing was, in other days at least, of greater than ordinary moment and evoked a greater than ordinary grief. Herodotus describes the obsequies of a Scythian chieftain. How the death cart was pulled from settlement to settlement. How the warriors gashed themselves and stabbed arrows through their hands (left hands, of course). Irving

Goldman describes the same sorts of royal corteges in Polynesia, often from islet to islet by canoe, and everywhere the same mortifications, the same bloody excesses of grief. Ritual grief, Dr. Goldman calls it, the emblem of deference to political power.

There was more. The death of kings in Tahiti and Tonga touched off orgies of killing, looting, destruction, to be ended only with the installation of the next king. Much the same violent interregnum followed hard on the royal death in certain African kingdoms. For the body of the king was more than common clay. It was as well the body social, the very flesh of order and civility. In this setting, social violence can be seen as a kind of mortification, an expression of grief, exponentially enlarged for him who was god on earth.

Pain and faith

What is proper and pleasing to earthly powers must be doubly so on high. The mortification of human flesh must be seen, in the celestial equation, as more than deference. It is an act of faith and confidence and even love. George Catlin's description of the Mandan O-kee-pa says it best. We saw through his eyes something of the great public ceremony: the masked dancers, the routing of the evil spirit, the triumph of "the Father of Buffaloes." We did not, however, enter the great lodge for the private rituals of faith which began at *her* order. This Catlin was allowed to witness because he had, on that day, painted a portrait of a priest so marvelously lifelike that the Mandan could refuse him nothing. He was even allowed to include friends and traveling companions, some of whom had been theretofore denied admittance.

Inside he saw the brave young men, that year's crop of "graduate warriors," as he called them. For four days they had abstained from food, water, and sleep and were lying weakly around the edges of the lodge. He watched as they were called, one by one, to the masked priest, who held a knife deliberately dulled. One by one the young men were gashed on calves and thighs, on upper and lower arms, on breast and back. Under each set of incisions, skewers were placed and to them tied the rawhide thongs by which the sufferers were hoisted to a wooden frame and there left to dangle. Dried buffalo skulls dragged on the leg and arm skewers. Then attendants spun the youths around and around. Only at this point did they cry aloud their prayers to the Great Spirit. After twenty minutes of twirling, they fainted (Catlin later painted the scene just at this point) and were lowered to the ground. There they remained. "They were," wrote Catlin, "here enjoying the inestimable privilege of voluntarily trusting their lives to the keeping of the Great Spirit, and chose to remain there until the Great Spirit gave them strength to get up and walk away."

When each was able to move, with his buffalo skulls still bouncing along behind, he went immediately to another masked priest by the door. After a brief prayer of thanksgiving, each laid the little finger of the left hand on a block, where it was struck off with a hatchet. Some, said Catlin, immediately offered as well the forefinger of the hand. Some, indeed, would in later life voluntarily undergo the whole ritual again, sacrificing the little finger of the right hand to their faith.

On emerging from the lodge, the exhausted candidates were pounced on by "assistants," who raced them or dragged them, fainting, around and around a large circular course until the dangling buffalo skulls were ripped free of their moorings, and only then were the braves allowed to collapse. Again each was left where he lay "until the Great Spirit gave him strength to rise upon his feet." Secular as well as spiritual eyes were interested, the chieftains ever alert to note who among the candidates best endured the trials.

Catlin saw some fifty young men test their faith in this ritual, from which the Sun Dance of the High Plains must surely derive. He saw none quail and none cry aloud except when this was permissible—at the spinning and the racing. Indeed, the youths, seeing Catlin's eyes upon them while their incisions were made, called him to look closely in their faces, smiling as the knife did its work. Catlin could not prevent his own eyes from filling with tears.

Faith and pain unfelt

The phenomenon of smiling pain has been reported in many circumstances by many observers, recently (in 1967) by the

Indian archaeologist D. D. Kosambi. He witnessed then the rite of hook-swinging, in many ways an analogue to the Mandan test of faith. It was once, he believes, a tribal custom in the Deccan, the southern plateau region, one long since assimilated to the dominant religious forms.

During the full moon of April, a young married man, member of a particular clan, impersonates the tutelary deity of the village. In the culminating rite, hooks are set in his lower back muscles. These are in turn attached to a long, cranelike structure, on the arm of which he is swung in giant circles.

Photographs reveal the devotee's gaiety throughout. And in the days following, the author saw no trace of infection in the wounds, rubbed initially with wood ashes. In two weeks, he added, the scars were barely visible.

Many field workers in West Africa have written at length—and indeed have filmed—the tribal use of judicial ordeals: the ordeal of the hot knife, of boiling oil, of liquid poison, which alike "catch" the guilty or perhaps those uncertain of themselves (the lie-detector test is a Western analogue). "The trials work for Africans," says Esther Warner, who was there, "because they have implicit faith in them."

Believing makes it so?

The power of belief is nowhere more dramatically enacted than in ritual firewalking. It appears (and is still practiced) in so many different parts of the world as to make one long to know more about the symbolism and the sacred significance of pedal mortification. Firewalking has been observed in Greece, Bulgaria, and Spain; in aboriginal Australia and among the Zuñi of our American Southwest; in the Pacific; in Japan; in India. Always observers have come away astonished at the absence of pain or evidence of burning. And always practical reasons are adduced: the feet are calloused, they sweat, they are coated beforehand with protective substances. One specialist suggested there might be a special manner of walking—the foot lowered firmly and in such a way as temporarily to quench the coal's glow. Practical answers, surely, but none of them demonstrably true.

Another practicial set of answers has recently been discovered by Solomon H. Snyder and Candace B. Pert of Johns Hopkins University. They have located within the body's nervous system natural substances very similar in composition to the opiates. They have also found the very localized receptors of these substances. The discovery is expected to shed light on the nature of analgesia and of addiction, and how both can better be managed. It also holds implications about the body's natural potential for dealing with pain. Yet how this potential can be summoned and under what circumstances is something that Western science has not yet learned. Perhaps self-hypnosis may be one way. But what about the fire? How explain the *absence* of lesions, blisters, and charring where, by rights, those ought to be even if the perception of pain itself can be effectively banished?

Perhaps there is something to belief after all.

In April 1972, anthropologist James M. Freeman witnessed a firewalk in village India. The ritual had been introduced some twenty years previously (tribal origins, claims Kosambi) to honor several local forms of the mother-goddess Kali. It centered on Khalasis, a low-caste man and woman, adept at entering the trance state and who were believed to be possessed by, became in trance, the living representatives of the goddess herself. Candidates for the firewalk bore considerable expense for the festival and underwent a ten-day program of rigid spiritual preparation which included fasting and abstinence. Those chosen by the goddess-surrogates would walk seven times over the twenty-yard length of glowing coals. They would, some of them, carry sick children or those dedicated by their parents to this oblation.

On the festival day, Dr. Freeman reports, one of the Khalasis was challenged by the local "magician," a man much feared but tolerated because of his ability to cure certain illnesses. He himself would walk, declared the magician, without prayer or preparation of any kind. He did walk, and lo! not even a blister rose to punish the blasphemy. But the Khalasi was himself burned and many others besides, too many to explain by way of impurity or the breaking of a taboo. Dr. Freeman attributes the mishap to nervous panic and to shaken faith. When doubt comes in the door, belief flies out the window, leaving only pain behind.

Therapeutic pain

Yet sometimes pain fully felt and fully appreciated is deemed necessary—the very medicine for life and health, to be taken without a teaspoon of sugar, and the nastier the better. Whipping in particular has been an accompaniment to medical treatment among many of the world's people. The "doctor" might whip himself to rev up his healing powers. He might instead whip the patient. Whipping has often been thought to frighten demons. And so it was that epileptics, in the old days in India, were often whipped. So were madmen in the London hospital founded by Sisters of the Star of Bethlehem and which, by the time of Henry VIII, had become the infamous Bedlam.

One finds the therapeutic whip plied in all sorts of circumstances. Cubeo boys of the upper Amazon region were periodically flogged by masked dancers. It was an experience quite different in intent from the ordeal of the stinging ants which tested a boy's courage or the nose full of stinging peppers with which pert and pubescent girls were given their comeuppance. ("You've been pleased to be disobedient, young lady. Let's see just how much pain you can take!")

A boy's floggings began when he reached his eighth year and were considered essential to growth, being called, as Dr. Goldman says, something very like "bodying." The boy was first held and whipped and then, as he approached manhood and all through life thereafter, was expected to accept whippings voluntarily for the sake of general well-being. Good and good for you!

Something of that feeling could be found, until recent years, in the school systems of England and America, where the paddle was seen (by authority figures if not students) as potent adjunct to learning rather more than as instrument of chatisement. Indeed, it is a view honored even in antiquity. "The schoolboy has ears on his back" was the schoolmaster's favorite proverb in the Mesopotamia of four thousand years ago. Ours preferred "Spare the rod and spoil the child" or, even better, "Whom the Lord loveth He chastiseth!"

Pain, fortitude, and power

The parents of the native North American child hoped and prayed and schemed to make him fit for a vision. Just as we scheme and sweat to send our children through good colleges. With the baccalaureate comes social rank and economic entrée.

With the vision came power, and with power came the good life. Contact with the spirit world, however, was possible only through the practice of austerities, through the mortification of the flesh. And even then there was no guarantee of passing marks. Parents therefore began the training process as soon as possible. Early in his childhood a boy learned to undergo with cheer the isolation in lonely places, the hard fasting, the offering of flesh or fingers so as to move the Powers to pity and patronage. Long before the arrival of puberty, children had thus been conditioned to accept the proper roads to self-realization in the way that our children, most of them, early accept the rigors of the academic life.

Yet it is in the specific rites of puberty that specialists have taken the warmest interest, particularly when those have involved genital surgery. Psychological interpretations of such rites have focused generally on the presumed envies of the opposite sex and of fears regarding copulation, death, defecation, and incest. One is hard put to say whether such fears and feelings might not be the result of such rites rather than the stimulus for them. "Besides," some critics insist, "one ought to distinguish between first and final reasons." Surely the recorded meanings of circumcision and clitoridectomy are nearly as varied as there are separate peoples. There are religious reasons, physiological reasons, and even beauty reasons.

As for first causes, without a time machine we are out of luck. Yet could we skim the world invisibly when it and our forebears were new, one suspects the answer would still be in the plural. Only the body itself provides the common denominator. It grows, it matures (with all the difficulty maturity entails), and as James L. Brain says, "It must be socially placed."

The body is, in fact, a pedagogical instrument, susceptible to the lessons of pain and at no time more so than when hormonal saps begin to rise. It was at this point fifty years ago that well-to-do American boys were often sent to military school, where their wild ebullience could be transformed with painful discipline into manly self-control.

But the lessons to be taught with pain are not everywhere the same and may never have been so. Certainly the control of sexual urges—the unrestrained exercise of which can be so destructive to the fragile social order—was one lesson to be taught. But there were others. Be steadfast, be brave, be self-reliant. Be aggressive; be restrained. Be the best; be like everyone else. Be industrious. Be generous. Be devout. Be spiritually in tune.

The lessons of pain among Australian Aborigines all focus on the Dream Time, that sacred "everywhen," and the Pitjanjara, visited by archaeologist Richard Gould, manage to extend them into what is virtually continuing education. The young man begins at puberty with circumcision. After that comes tooth removal. Then nose-piercing and, in slow succession, firewalking, chest scarification, and subincision. The scars from this last operation will be periodically reopened throughout his life. For each new trial of pain, the candidate retires from camp in order to learn the associated sacred lore. Through visits to strange folk and many cult centers, his knowledge grows until, stricken with years, he departs this earth in gestural communion with his clan's totemic being.

Pain, penance, and conquest

The pains of penance in a world bounded by sin are not only to be endured bravely, they are to be courted. In such a world the self is seen as an uneasy alliance of soul and body, the latter an ever-present threat to the common enterprise, which is glory or surcease in Heaven. Like an overgrown child, the body is naturally gluttonous, slothful, concupiscent—its carnal nature prone to error which only vigilance and punishment can curb. It is not a pedagogical instrument to be taught with pain. It is not a power instrument to be honed with pain. It is an animal to be subdued and discarded as soon as God's tasks have been fulfilled.

Among only three peoples of the world can one find that passion for penitence which finds expression in mortification. One finds it among ancient Mesoamericans. One finds it in India and among those influenced by the various creeds nurtured in India. One finds it in early Christianity up to the time of the Renaissance.

Perhaps it was the suffering of Jesus and the terrible manner of his death that initially prompted emulation among the devout.

Perhaps it was the Roman persecutions and the agonies of martyrs that drove guilty survivors to self-punishment. Whatever the stimulus, extremes of mortification became the heroism most admired, though infrequently imitated.

Consider Saint Simeon Stylites, the Hermit, born before A.D. 460 in what is now Lebanon. Accepted into the brotherhood of a local monastery, he at once irritated the monks with his champion austerities, and they complained to the abbot. On one occasion Simeon appropriated the well rope, winding it so tightly around his body from neck to loins that it grew into his flesh as the bark of a tree incorporates a chain. At length the abbot gave the young man permission (with some relief, it must be supposed) to remove himself to a mountain den. After that, in flight from the importunings of his growing public (or perhaps to offer a better view), he made his home on a series of ever-taller pillars in the desert, where he became a prominent tourist attraction and the cause of mass conversions. As the Reverend S. Baring-Gould in his *Lives of the Saints* assures us:

> There the fame of [his] austerities wrought upon the wild Arab tribes and effected what no missionaries had been able, as yet, to perform.

In those days penance gripped the imagination, and austerities marked the hero-journey of saints, even as in India. By the twelfth century, mortifications were of a less flamboyant sort. Saint Thomas à Becket—inflexible antagonist of the second English Henry—became as Archbishop of Canterbury a secret ascetic. He had a grand and comfortable bedstead but slept on the floor. He treated dinner guests to sumptuous repasts while taking nothing for his own appetitie. And he practiced assiduously the technique of mortification most popular among saints, self-scourging. Sometimes he ordered his chaplain to ply the whip. The resulting stripes he covered with the coarse, plain habit of a monk hidden under the elegant episcopal robes. Only after his assassination was the full extent of his discomfort revealed: under the monk's garb he had worn a hair shirt and hair breeches as well. Both were "boiling over" with lice, a fact which struck the populace with awe and was the source of many legends regarding the fortuitous commutation of vermin into pearls.

By the sixteenth century the ascetic no longer had instant claim upon popular reverence, and the great Ignatius Loyola, Spanish founder of the Society of Jesus, was once given a drubbing by local playboys. Saint Ignatius himself had found no peace in his austerities, in self-starvation, in the vigils, the scourgings, the pain. Only in a life actively devoted to preaching, teaching, and study did he find redemption.

Even Saint Francis of Assisi, the most devoted ascetic of them all, was heard at life's end to regret having offended his brother the body and given it pain.

Pain conquered and courted anew

In today's world we have the means to banish nearly every unwelcome pain—in illness, dentistry, injury, and medical procedures of every sort. And we account this the greatest of blessings. No wonder we are shocked by the person who mortifies his body without need, without cause. Now it is true that excesses of pain can carry the seeds of pleasure and exquisitely sensuous experience. This, after all, is what masochism is all about. But to so label the courting of pain in other cultures is to close the door to understanding. We need not believe to accept belief and to appreciate its power.

Besides, who says we don't mortify the flesh?

There are, of course, few initiation rites for our maturing young other than those circumspect denominational services which mark, rather, the entry into adolescence than into adulthood. Nobody ever says, "*Today* you are a woman. *Today* you are a man." Adolescents experience all alone the biological turning points and then the early sexual encounters into which they are now permitted to fumble without sanction or ceremonial recognition. There are no tests, no diplomas to give maturity real worth. The acquisition of the driver's license is a minor event, and the college degree often seems less a conquest than a contest of endurance.

No wonder the young, especially young men, evolve their own informal passage rites. From fraternity hazings to drag races with the family car; from drug dares to gang rumbles—all are tests. How much can you take without turning chicken? What are the limits of courage, daring, confidence, strength? Keep up with your peers and *they* will say, though society may not, "Today you are a man!"

The devotee of tennis anticipates and appears even to welcome the concomitant assaults to elbow and wrist, and suffers willingly to improve the game. The long-distance runner, who has supplanted the cyclist as the driver's most common hazard on the road, willingly endures the pain, the exhaustion, and the injuries which are the inevitable accompaniment to this pursuit. For there is in running—so we are told by runners themselves—an element of mystic communion between mind and body, even a sense of pure connection with All There Is. It appears to be a feeling not unlike that which Tarahumara Indians of western Mexico associate with running. At first a hunting method, it became for them, as for so many other tribes of the American desert regions, a religious exercise.

For the end of running is the testing. How far beyond its limits of exhaustion, its limits of oxygen deficit, can the body be driven or cajoled? How much pain can it endure without whining or collapse? And what is gained thereby?

Why, a triumph is gained. A conquest of the flesh. For some this may be life's only victory.

11

USES OF THE FLESH

Fellow Feeling, Fellow Feeding

What is the price of a human body? Of what value the quarts of blood, the bones, the pounds of flesh? It used to be reckoned at just above four dollars. All the books of physiology told one so. Now, what with inflation and a resurgent interest in trace metals, the price has escalated considerably. Several thousand dollars is the announced going rate.

And what of life that animates the flesh? Cheap or priceless, it would seem, depending on whose life is in question and the size of the emotional stake vested therein. A life loved and freely offered is the ultimate gift, the gift which demands attention from the Powers of the universe and which must be gratefully received. Purity and pain, self-imposed and self-inflicted, may stir the pity of the gods, but the gift of a life must be recompensed, and that promptly. Such has been the belief among ancient humankind and, in various out-of-the-way places (and some not so out-of-the-way), may be so to this day. Only let some great catastrophe occur, some dire upheaval, and frightened survivors will seek a scapegoat to punish and martyrs to honor and sacrifice.

The offer that cannot be refused

The Hebrew word for sacrifice means "gift," and in early times the gift required was often the ultimate one. We read in the Book of Genesis how God tempted Abraham, saying,

> Take now thy son, thine only son Isaac, whom thou lovest, and get thee into the land of Moriah; and offer him there for a burnt offering upon one of the mountains which I will tell thee of.

And faithful Abraham took Isaac, the son got in the very winter of his long life, and did as he had been commanded. Only when his knife was poised to slay did the Lord intervene, saying, ". . . for now I know that thou fearest God, seeing thou hast not withheld thy son, thine only son, from me."

Such a test of piety may have been uncommon in Abraham's time (perhaps 1800 B.C., perhaps much earlier). By the time of the exodus out of Egypt, coastal Canaanites had made infant sacrifice a religious commonplace. From the Book of Leviticus on, the Children of Israel were warned incessantly against "making your sons to pass through the fire," against "giving [your] seed to Molech." This may refer to the idol in whose furnace-belly the little offerings were burned or the mode of sacrifice itself. Biblical injunctions were not, it seems, altogether effective. At least two kings of Judah ostentatiously offered their own children in this way. It was a custom taken to Carthage, that largest of the Phoenician colonies, and perpetuated there long enough to shock and outrage the enemy in Rome.

The Carthaginians were not the only nation of the time to favor children as sacrificial gifts. The Persians preferred their religious thank offerings to be very young, seven to fourteen in number, and buried alive. Half a world and near two millennia away, the Inca also dug their child sacrifices into the ground, enclosed in stone cists. From a high Chilean mountain, one of these offerings was recovered in the 1950s. Frozen in his cist for five hundred years, he is today as he was in death, a little boy of eight or so, his head resting on his updrawn knees, his face peacefully composed in sleep. Mercifully, he had been drugged before interment, so that death caught him unawares. For all the little Peruvian victims such care may have been taken. The Powers, it seems, extract consent and grant their boons when the gift is as willing as the giver—or can be made to seem so.

Giving graciously

Consider Jephthah the Gileadite, a mighty man of valor in Israel who, as a price of victory, vowed to give the Lord the first being who greeted him on his triumphal return. That person was his daughter, his only child.

> And she said unto him, My father, if thou hast opened thy mouth unto the Lord, do to me according to that which hath proceedeth out of thy mouth.

This time there was no heavenly intervention. The gift went consenting to her fate.

We read consent in the face of Tollund Man, turned to leather long since by the acids of a Danish bog. The cord with which he was strangled or by which hanged is still about his neck. Apart from a belt and a little cap he is naked. Perhaps he was an offering to the Earth Goddess whose image cart he once pulled. Not quite two thousand years ago, the Roman historian Tacitus described religious processions of this Teutonic divinity and what happened to her devotees: strangled and bog-buried. Some may have struggled and protested; Tacitus does not say. Not Tollund Man, however, who closed his eyes in perfect acceptance of his fate and perfect confidence in what was to come thereafter. His people doubtless considered his devotion commonplace. Our verdict would be brainwashing. And who is to say which is right?

We cannot know when human sacrifice began to represent

the ultimate gift. Remembering the belief (strong among so many hunting-gathering folk) that animals themselves are consenting victims, remembering the art of the Ice Age cave with its preoccupation with animal images and themes, one might with reason seek a beginning in these early times. The felt burden of responsibility is plain: life for life, gift for gift. There are several clear representations of sacrifice, among them the head and fleshless spine of a bison, a horse gushing blood. And always animal victims are surrounded by human forms in attitudes of reverence.

Yet in the art of the caves there can be found no unequivocal evidence of human sacrifice. Nor do we find it among hunting-gathering folk of more recent times. Not at least among the roaming ones, the ones for whom struggle was life's law. Only more prosperous and populous herders or farmers, more often than not, could afford the human offerings. The more complex the social setting, the greater the tab. And in some ways, the more easily paid. One need no longer sacrifice one's own. There were expendables at hand: battle captives, criminals, slaves. The Carthaginians themselves had taken to substituting for their precious offspring the infants of slaves or even baby animals. Until the Roman danger was at hand, and counterfeit sacrifice was blamed for the catastrophe.

Even with slave surrogates, the fiction of consent was everywhere carefully maintained. Victims were promised, or believed they could achieve, immediate entry to a better world, and went willingly to the executioner. But not always and not all. There is a story (probably apocryphal) told about one of the cities of the Yoruba tribe, now part of modern Nigeria. In older times great occasions of the king's reign and of his death were celebrated with human sacrifice. Before losing their heads, victims were given oral messages to be delivered to the royal ancestors, surely by now consorting on high with fellow gods. And then one year the scenario changed abruptly. A prospective victim refused to cooperate. He would not, he declared, play postman on the other side. And in ringing, defiant tones he dared to curse both king and executioner. Gasps and consternation! No omen could be more terrible. What to do? Obviously, release the man and find another messenger. And from that day until the sacrifices lapsed, human victims were gagged to forestall cursing. You can see the clay and bronze heads of just such victims displayed in the Royal Museum at Ife, holy city of the Yoruba.

You can take it with you

When an ancient Oni of Ife considered his own demise, he surely planned to keep up appearances with a ghostly retinue of suitable size and elegance. The rulers of archaic kingdoms the world over (and some not so archaic) have firmly believed that you *can* take it with you. According to Herodotus, Scythian kings liked the company of the royal guard. Some fifty of these, therefore, impaled on their horses, were arranged about the burial mound, there to present eternal arms. The evidence of Scythian tombs excavated so far, together with the ethnographic record, has confirmed so many of the Greek historian's observations that there seems no reason to doubt him in this.

Archaeology has also uncovered the mastabas of First Dynasty Egypt—before royalty took to substituting pottery figurines for the real thing—and found them chockablock with servitors of every ilk and occupation. In preconquest Mexico, so we learn from Dr. Coe, a dead king's wives and courtiers were slain and separately cremated. Among companions specifically mentioned for Tarascan royal funerals are: the woman who carries the royal lip plugs, the woman keeper of the finger bowls, the woman keeper of the royal urinal, the fan-bearer, the sandal-bearer, the bearer of war jackets. The lists of occupations is often strange to us but does include at least one recognizable figure: the doctor who failed to effect a cure. The entire cast of characters is said to have been made drunk and then beaten to death.

Shang dynasts with a strong emphasis on quantity included in the royal tomb large numbers of captured "shepherd folk" who could augment the king's claim to immortality with the essence of their little lives. The great Ch'in Shih Huang Ti, whose line came to an end in 206 B.C., bringing the dynasts of Han to power, is said to have taken regiments of victims into his tomb. That tomb has recently been uncovered. It does indeed contain regiments of soldiers and their horses, but they are all of clay, life-size, and beautifully crafted. Even tyrants at some time must reckon with lapses of faith.

That royal companions-in-death often did make a voluntary exit was demonstrated by Sir Leonard Woolley in Mesopotamia. There in 1927 he excavated the Royal Cemetery of Ur, constructed nearly five thousand years ago for the ruling family of that city. In one of the huge pits he found the remains of over seventy courtiers. Having followed the royal bier, they had apparently settled themselves sedately and ceremoniously around the death house. Royal carts and draft animals were there together with their grooms and drivers, each in his appointed place. Musicians were there with harps and cymbals—which, thinks Woolley, they may have played right to the end of the drama and their own lives. There were ladies-in-waiting wearing jeweled headdresses. There were soldiers in their battle gear. At a signal each held out a little cup of clay or stone or metal, which someone filled, perhaps from the giant copper vessel Woolley found in one grave. They all drank the poison and then, calmly lying down, composed themselves for death. Later, before each pit was filled with earth, someone came down to put to rights any unseemliness occasioned by the throes of dying.

Barbarous, you say? Simpleminded, you say? Herd behavior, impossible for the modern sophisticate? Wrong. Not so very long ago an American cult leader not only led his followers in exodus to Guyana but also into death. Like servants of the court at Ur, they obediently drank poison and obediently laid themselves down to die. Surely the Reverend James Jones went wherever he was going with the kind of ghostly retinue a Sumerian king might well have envied. Some nine hundred souls accompanied him in death. Most of them willingly and with consent.

Death along the Mississippi

Of that drama of extinction some holdouts remained alive to tell the tale. Of older immolations there are few testaments. Eyewitness accounts are still more scarce. One such was recorded by Sieur Antoine Le Page du Pratz, who came to the lower Mississippi in 1719 with a group of French colonists. He soon became a friend of the Natchez, the tribe which dominated the region.

Such a favored friend, in fact, that he was allowed to witness the obsequies of Tattooed Serpent, much beloved brother and war chief of the king, the Great Sun. Now the Natchez had at the time what must surely be the world's most curious system of class division. An aristocratic gentleman's own children invariably ranked one step lower than himself, for social position was inherited through the mother. Generation by generation, some of the Natchez sank to the stinkard class, from which, by and by, their descendants could then expect to rise. This was because stinkards were sought in marriage by members of the three upper classes. There was a very good reason. The aristocratic spouse could not be expected to accompany one in death. The stinkard had to go along. Such was the condition of upward mobility.

Tattooed Serpent, by virtue of rank, could command a large entourage, and did. His stinkard wife, elegant of bearing, was proud to go. So was a noblewoman who volunteered, together with sundry servitors and several old women substituting for their stinkard relative, one who had earlier avoided playing escort to his dead wife (a lady of high degree) by taking refuge with the French.

Each leading character was attended by eight kinsmen who would perform the execution (achieving social promotion thereby), and the entire cast was carefully rehearsed. Then came the performance. As the funeral cortege wound in slow circles from house to temple, du Pratz saw a stinkard couple repeatedly throw the corpse of their strangled infant under the passing bier. By this act they too moved up the class ladder. At the place of departure, the companions-in-death were fed stupefying balls of tobacco and then, says du Pratz, strangled so expeditiously that one hardly realized the deed had been done.

The Natchez were familiar with even lengthier dramas of death, performed without benefit of analgesia. So were most of the tribal groups in the eastern part of North America from the Gulf to the Great Lakes. For it was customary there to put war captives to death as slowly as possible and with the greatest possible infliction of torment. It seems as if pain itself had become a kind of offering, a gift all wrapped up in battle dress.

The sacrament of battle

Nowhere in North America was this complex of torture and sacrifice more elaborate than among the Iroquoian-speaking people of the Great Lakes region. To bring a captive back home successfully was the greatest honor for a Huron warrior, a Seneca, a Mohawk, or any of the other related tribes of the region. Though his bow finger was immediately bitten off, the captive entertained some hope of survival through permanent adoption by an enemy family. If they liked his looks and if he had not been maimed in battle, he might be allowed to replace a dead family member. Temporary adoption there would be in any case, but the hospitable and charming adoptive family could at once or at any time later consign him to the torture, and themselves apply knife and brand with greatest glee. It is in this fiction of adoption that one can read the gift: the family's faith offering of its own.

And there is more, as ethnohistorian Bruce Trigger demonstrates. Throughout the period of torment—which might be prolonged for days—the men and women of the village had to remain watchful and sexually continent. Never must the captive be allowed to die before dawn, before he could be dispatched on a high scaffold and in the eye of the sun. If he had borne himself bravely throughout, defiantly singing his war song and daring his tormentors to even greater excesses of cruelty, his heart would be offered and consumed. His flesh, cut by cut, had already (and in his hearing) been bespoken and would now be butchered and cooked.

In this ritual application of pain and ritual communion of the flesh—customs which appear to have begun among the Iroquoians sometime before A.D. 1400—we can read the effect of influences traveling steadily northward from Mesoamerica. It cannot be said just when ritual torment and ritual cannibalism began there. Human bones are among the most common remains of Olmec diet to be found in the sealed garbage dumps of San Lorenzo. And one of the most striking themes in the Maya pattern is that of human sacrifice, of ritual beheadings and ritual torture, of victims bound to the scaffold. These are scenes which rarely appear on the great monuments of Classic times but which, as Michael Coe has shown, are commonly displayed on funerary ceramics.

A Mesoamerican holocaust

From these and from other predecessors in Mesoamerica, the Aztecs inherited ritual forms. They then gave the sum exponential multiplication. The number of human victims sacrificed annually in Aztec Mexico has been estimated at 250,000—1,000 to 3,000 for each of the one hundred large temples in the empire and proportionately smaller numbers for the less important shrines. And whence came the victims? In the main from neighboring kingdoms which themselves sought Aztec victims in what were called "flowery wars." These were designed ultimately to provide temple offerings and high honor for the warriors who supplied them. The captive fully expected courteous treatment before his death. He might even be chosen to impersonate a god in ritual drama so movingly solemn that spectators wept and cried aloud when, at the pinnacle of the steep temple pyramid, his heart was wrenched from his living bosom and offered to the sun. It was on this sacrifice, this offering of pain and blood, that the gods depended for their continued life. The victim was himself comforted at the moment of death by the certain knowledge that with his brave bearing he had achieved instant transport to paradise. There was more to the gift than courage. There was the flesh to be consumed.

> Every day we saw sacrificed before us three, four, or five Indians whose hearts were offered to their idols and their blood plastered on the walls, and the feet, arms, and legs of the victims were cut off and eaten, just as in our country we buy beef from the butcher.

So wrote Bernal Díaz del Castillo, a Spanish soldier in the expeditionary army of Cortes. There is little reason to doubt his observation (though hindsight may have lent some exaggeration). The Aztecs themselves painted such scenes for the monk-ethnographer Bernardino de Sahagún. And archaeology has uncovered near the temple at Tlateloco stacks of buried rib cages minus appendages. Heads were usually exhibited on skull racks, and at least one limb was given to the victim's captor.

If there was much sacrifice in Mexico, there was also a considerable disposal problem, which the Aztec solved by internalizing the remains. It was an act meant to symbolize communion with the gods, becoming one with them. This pious yearning, though certainly not the reality, is well known to Catholics in the most sacred moment of the mass: "Take ye and eat, for this is my body."

Hidden hungers and cannibalism

The Aztec ritual, Michael Harner contends, was just an elaborate excuse to procure meat, even if only soldiery and the elite had better meals thereby. The poor could still hope for advancement through military service and the diet that went along with it. Why not? says Dr. Harner. There was little enough protein and fat to go around. Game had long since been depleted, and there were no large domestic herbivores available. Worms and algae skimmed from lakes are not really all that palatable. The staple corn and beans would have had to be consumed in huge quantities to provide protein sufficient for tissue repair as well as for energy. And crop failures were common. Hidden hungers will out, says Dr. Harner, and in the light of that hunger the several independent Mexican kingdoms, both within and without the Aztec empire, could be regarded as elaborate game preserves, to be raided at leisure.

Bernard R. Ortiz de Montellano, replying in *Science*, questions Dr. Harner's dismal view of Aztec nutritional resources. Leaching corn in alkali—a process well known to most American Indians (we call the finished product hominy)—enhances its amino-acid content. In Mexico, he says, there was no pellagra, that deficiency disease common to those dependent on a plain corn diet. If deer were few, people could and did eat waterfowl, armadillo, gopher, weasel, mice, snakes, dogs, plus worms and insects of every sort. Yes, there were indeed famines—three big ones are recorded. Bodies are said to have lain unburied where they fell. But no one ate them. Why not? And why was the greatest sacrificial toll taken (in thanks) at the harvest seasons and not (for meat) when supplies were at lowest ebb?

Never underestimate the power of religion, admonishes Dr. Ortiz, nor the single-minded, even rabid, zeal of people in the clutch of an idea. (Remember Jonestown, the Nazi Holocaust, the Inquisition—all sacrificial or judgmental uses of the flesh; none explicable in pragmatic terms.)

Yet surely the knife of explanation can cut more ways than one. Surely the expression of a cultural pattern can be permitted the ambivalences we tolerate in the single personality. Flesh available to the diet may be rejected out of hand because it is ritually unclean, and never mind the hidden hungers. Flesh intended quite devotionally for the gods need not be despised on the table. People, it seems, can say what they mean and mean what they say. People can also say one thing and do something else altogether.

The cultivated taste

The Tupinamba, who lived around the coastal hump of Brazil, *said* they ate their enemies both to insult them and to honor them. And certainly there was a sacred component, since the captive (and ritually dangerous) jaguar was dispatched with the same ceremony accorded a human victim. But they also liked the taste. So we learn from the sixteenth-century writings of European captives who somehow managed to escape the pot. In a context of intensive farming and rich resources of river and of sea, human meat was quite simply preferred to anything else. In order to guarantee supply, the Tupinamba encouraged the captive man or woman to breed—much as the shepherd breeds lambs for the slaughter.

The Maori had brought the habit of cannibalism with them to New Zealand from their homelands in eastern Polynesia, and not very much else to bolster the diet. With a protein base consisting largely of fish, dogs, rats, and birds, they were just as appreciative of human meat as the Tupinamba and, because of real need, correspondingly more forthright about saying so. One ate enemies as a matter of course. One also ate slaves and occasionally, during the privations of a siege, the youngest children. (Families did save feelings by effecting an exchange.) Every great occasion—the tattooing of a young woman of rank, for example, the lifting of mourning, the visit of a distinguished chief—required a human entrée. And if there was no slave available, family members themselves might draw lots in order to preserve the honor of the kitchen.

In Africa several tribes had notorious reputations for cannibalism, among them the Azande, who claimed that, in a pinch, they might eat chimpanzee, which was almost as good as a man. But that was long, long ago, generations ago, and there are no longer in Africa, or Polynesia, for that matter, old men who insult one another by pointing to their own teeth—an eloquent gesture meaning "I've chewed your folks!"

Only in New Guinea and some of the Melanesian islands is the memory still green. It was just ten years ago, says Klaus-Friedrich Koch, that two missionaries were eaten by the Jale, warlike neighbors of the Dugum Dani in West Irian. Some Melanesians have assured Western visitors that the meat of white men really stinks and would never be eaten by them. The Jale apparently did not cherish such reservations.

The ultimate memorial

Again, wherever man eats man in New Guinea (the habit is now thought to be extinct, though not certainly so), the rationale is generally war, the rituals of war, the rituals of revenge, the rituals of courage honored and, with the flesh, internalized. They also say, "Man tastes as good as pork, if not better."

The eastern Fore, late converts to the habit, told Ronald Berndt in 1951, "This is sweet. Are we mad? Here is good food, and we have neglected it." There was no mention of internalizing good qualities, or any mystic communion, for that matter. Berndt claims a shortage of pigs among the Fore. Whatever the underlying reasons, they very quickly learned to eat not only enemies but also their own relatives. Not all their relatives. One did not eat one's parents, and vice versa. In fact, one paid with pigs those who did dispose of Mother's remains. A husband did not eat his wife, but she relished every part of him, including the contents of his lower intestines. Loved ones on the point of death were said to direct survivors to eat their flesh (and sometimes even to dictate the cooking procedure) so that the crops would grow well. And as a further suggestion of fertility offering, relatives were butchered and eaten in the family gardens. But the Fore were not above hastening a sick person's exit or even occasionally beginning his dismemberment before death had been fully consummated. Their taste for brains, at least according to some specialists, figured in the occurrence of a fatal neurological illness called by the Fore "sorcery" and known to modern medicine as kuru. Robert Glasse, who studied this phenomenon, connected the high incidence of kuru among women to the fewer cannibal restrictions applicable to them (men considered female flesh weakening). Children under six, who ate anything their mothers gave them, were also frequent sufferers.

Other people have practiced familial cannibalism with rather more reverence or more horror or both. The entombment of a loved one within the alimentary canal may be considered a sacred duty or even a test of official position, but one should not enjoy it. Heirs of some African kingdoms were required to eat the hearts of their predecessors. A Chinese of early dynastic times who aspired to a dead man's position was required to consume his flesh—though this may have been restricted to some small part thereof. The founder of the Han Dynasty is reported by Ssu-ma Ch'ien to have drunk the broth of his father. And the record is riddled with stories of young men who, lacking the will for this duty, were themselves killed or excluded from the line.

Among many tribes of the Amazon Basin, endocannibalism is reported, whether the requirement involves fresh meat or the ashes of cremation well cooked into the staple food. Some Australian tribes have practiced endocannibalism and many other death customs as well, most of them, like fellow feeding, designed to remove perishable flesh and expose imperishable bones, which can then be reverently installed in yet another resting place.

Dual immortality

The transformation process often involves what the specialist calls secondary burial, and we have clear archaeological evidence of secondary burial going back as far as the Mesolithic in the Old World and the comparable Archaic in the New (both beginning around 8000 B.C.). Some specialists have attributed this habit to the ancient belief that there's no place like home. The body of the traveler killed on a journey cannot conveniently be returned by his companions. But they can give him temporary burial just until the flesh has decayed. They can then bundle the bones and bring them back to be reinterred in the family vault, so to speak—or, more properly, under the living-room floor.

And yet there is also evidence both archaeological and ethnographic of bodies deliberately excarnated at home. The wall paintings at Çatal Hüyük clearly suggest that the huge Anatolian vulture was at least one agent of excarnation. One is reminded of the Parsi in India, who expose their dead on "towers of silence," where their flesh feeds the birds. Or of Tibetans, who also have offered the flesh of loved ones to the birds of the air and the beasts of the hills.

Many tribes of North America have practiced bundle burial. The Mandan retrieved the skulls of such burials and placed them in circles on the prairie, all facing inward toward skulls of the buffalo. Both the Huron of Canada and the Inca of Peru once yearly brought out the skeletons of the dear departed, dressed them in new clothes, and honored them at a great feast.

The difficulties of inhumation when the ground is hard and tools of excavation few is seldom suggested as one possible origin of secondary burial. Certainly the record is full of burials—primary or otherwise—in the easily worked debris of the communal dump heap.

Very shallow primary graves located near the men's house are dug by the Bororo, a people of the Brazilian Mato Grosso. For something over a month the corpse remains there to be honored by painted dancers in animal costume, to be avenged by the killing of a jaguar, and to be periodically drenched with water and palm wine. At last the bones are retrieved, washed, painted, and given final burial in a nearby swamp.

Remembering how often and in how many cultures the prescribed period of deep mourning is forty days, one cannot but wonder whether it is related at all to the time required for excarnation to be complete.

In 1907 ethnologist Robert Hertz synthesized in a famous essay the travel accounts of burial customs favored in central Borneo and among other peoples of the Malay archipelago. As Peter A. Metcalf tells us in 1978, very similar customs prevail in modern Borneo. The same is true in Sulawesi among the Toradja, who have even learned to tolerate the presence of photographers at such events. There the body is retained under many rich coverings for the period of decomposition. Mourners symbolically share the impure condition of the corpse, signifying this by eating the fluids of decay and by restricting as much as possible their normal diets. They are also silent. (This particular burden was, once upon a time, lifted with the taking of an enemy head.) When the bones are purified and ready for final interment in the family cave, together with handsome effigies of their living form, a great feast is given. In the old days a slave, magically deprived of his soul, was chained to a post and tortured by male relatives of the deceased. And the more cruel the torment, the greater the joy of the departed shade, whose spirit translation was thus made complete.

Hertz believed that the association of decay with the soul's journey to paradise was widespread. It was, and it is. A curious analogue of the Malaysian ritual is still practiced on some Greek islands where, says Margaret E. Kenna, loved ones are buried in the village cemetery for three years, at the end of which time the string of memorial feasts designed to procure God's forgiveness will have propelled the deceased into Heaven. His bones are then exhumed and placed in a family ossuary among the skeletal remains of his ancestors.

In the view of Dr. Hertz, cremation was really a process by which the bones could be more speedily released. Afterward, they and the purified ashes could receive reverential care. Mummification, he felt, by rendering the flesh bonelike, was yet another way to deal with its fickle corruptibility.

Unlike flesh, bones are pure, white, impervious to decay. They are imperishable. Or so they appear to the generation in which there lingers a memory of the dead. No wonder, then, to find them so often associated with that which is imperishable in the human spirit and which is remembered. But bones and spirit are anonymous. For the living they may represent power and force. But the force is impersonal and foreign.

It is in flesh, with all its penalties and pains, its susceptibility to ailment and its certain decay, that spirit dwells. It is through flesh that spirit can reach out. Flesh touches flesh. Flesh warms flesh. Flesh speaks and sings. Like a leaf it withers and, withering, loves.

> This thou perceiv'st [said the Bard], which
> makes thy love more strong,
> To love that well which thou must leave ere long.

12

THROUGH FLESH TO FANTASY

Sacred Senses

For three days the sick young man had lain on the mountaintop. Naked and painted white with clay, huddling under his buffalo robe, he had wept and prayed that something would take pity on him and heal his swollen throat. For three days he had been without food or water, and as darkness approached he lay down to wait, tense and hopeful. Suddenly he heard a rasping, whispery sound that became almost but not quite words, as if something owl-like wanted to be heard and understood. From the corner of his eye, the young man saw a tall shadow which seemed to take on human shape and spoke in a human voice, saying, "Look at me." But lo, there were only human bones where the shadow had been, all in a heap. And the young man slept and on the fourth night was healed. Owls had come to him and snakes and the small turtle with a long tail like storm clouds with a dragging train. Through these beings the young man was healed, and from them came his power to heal others.

Gilbert McAllister has told this story of the great Kiowa-Apache shaman Dävéko, who died in 1897 after a life of renown in the southern Plains. The visions which set upon his powers the seal of authenticity were eagerly, plaintively (though sometimes vainly) desired by all young men of his time and culture and were sought, as we have already seen, through reverence, clean living, and self-denial.

If you or I were to starve ourselves for days while keeping remorseless vigil, we too might expect to receive visions. Hallucinations, the psychiatric specialists of our own world would call them; false perceptions. In common terms, "The little man who wasn't there." Fatigue, emotional stress, fever, sensory deprivation or overload, migraine headache—these too can induce the trancelike states in which such "little men" abound. So can strong, insistent, repetitive rhythms—which may help to explain the popularity of the discotheque and its progenitor, the dance hall.

Bushmen of the Kalahari Desert dance. Among the last of the world's hunting-gathering people, these small men and women dance for hours and hours through the night to heat up the medicine stored in the pits of their stomachs. They dance in order to see people (as Richard Borshay Lee translates) like very small birds, and the world spinning, spinning, and the trees going round. They dance in order to touch their friends and, touching, cure. To the persistent, numbing rhythms, Dr. Lee adds as trance inducements the night, the fire, hyperventilation, and the concentrated will to believe. For very many people, adept at opening doors between this reality and the next, such ingredients have been enough.

But there are shortcuts to that other reality, to a world where one sees gods and heroes face-to-face. There are plants whose essences, internalized, made one with flesh, open the door quickly and without effort. Dävéko had one of these, though he didn't know it. He had tobacco. He smoked as often as he prayed, for smoke and prayer were one. And who can say for sure how much nicotine had to do with the pile of bones that wasn't there?

The all-American weed

Ordinarily classed as a stimulant, very strong tobacco can conjure up visions with the best of the hallucinogens. Or so some psychologists and anthropologists believe. And why not? Nicotiana belongs in the plant family which includes deadly nightshade and jimsonweed as well as the common potato. Certainly tobacco has been used as psychotrope in the Americas, alone or in concert with the more obvious mindbenders. (One doubts the Surgeon General's warning on cigarette packages encompasses just this aspect of tobacco.)

Of the many plants native to the Americas—useful, useless, or in-between—none has had a greater impact on the world at large than tobacco. And none has undergone thereby more change in its original matrix of meaning. Native Americans appreciated, to a greater or lesser degree, the physical effects of the plant. (Some, indeed, were and are as addicted as the modern chain-smoker trying to kick the habit.) It was, nevertheless, the sacred significance of use that counted and, in some out-of-the-way places in the Americas, counts still. Among the Andean Campa, for example, the word for "priest" includes the concept of tobacco.

Nicotiana grows wild everywhere in the hemisphere and in many species. At least two of these (one much stronger than the other) appear to have been domesticated in South America and diffused slowly north. Tobacco leaves, wild or tame, have been macerated and eaten as a soup thickened with manioc starch. They have been liquefied and drunk. They have been chewed with shell lime—a practice, found among some tribes in California and the Pacific Northwest, which brings to mind both the betel chewing (with lime) popular in Southeast Asia and the ever-present coca quid of the Andean highlander. Coca, the "divine leaf" of the Inca (for us the source of cocaine), was and is chewed to allay hunger, thirst, fatigue, and the pains of altitude sickness. And it is chewed with lime made from plant ashes compressed handily into a little gray packet.

But in the main tobacco has been smoked: rolled into cigars, shredded for pipes, burned as incense. And it is as a source of fragrant smoke that its use may well have begun. Native Americans have always loved the sweet savors of burning cedar, of aromatic copal, and of sumac. Before the arrival of domesticated tobacco in North America, various barks and leaves were even smoked in pipes. Smoke, after all, rises as the birds rise to heaven. Smoke coalesces as clouds gather in promise of rain. Smoke is therefore prayer and purity and plenty. These original connotations clung to the leaf in whatever form it was afterward employed.

Smoke cures. Among very many people of this hemisphere, it was the only native therapy at the shaman's command. He blew smoke over the sick person, fumigating him, censing him with sacred airs and breezes.

Smoke exorcises. Among the Cubeo, for example, smoke wards off the dangers of the river and the dangers of food for a young family newly emerging from the seclusions and restrictions of birth.

Most of all, smoke binds. In the days of great chiefdoms and confederacies, it confirmed the chief in office and bound him to duty. And among ordinary folk, the shared cigar enhanced affection, the proffered pipe amended all.

"Smoke the peace pipe" is a phrase that has worked its way into our litany of phrases. So has "Bury the hatchet." Native Americans, says ethnohistorian Robert Hall, combined the two actions in a single artifact.

He arrived at this conclusion by way of questions. Why was it, he wondered, that among Indians of historic times the pipe stem, the so-called calumet, was so long and narrow? Why was this stem more sacred than the bowl? And why, among certain prairie folk, was this instrument of peace made to look so much like an arrow, with its fletched feathers and military ornaments? Well, he thought, if swords can be beaten into plowshares, why can't arrows be transformed into pipes? The writings of French explorers of the sixteenth century indicate that an exchange of weapons among Indian groups did signify friendly intentions.

But, Dr. Hall remembered, arrows were latecomers to the North American arsenal, arriving in the eastern regions not much before A.D. 400. Nor do arrow-shaped calumets enjoy notable antiquity. The earliest pipe shapes are those in the Mississippi region, dating perhaps from 1000 B.C. Tubular, elegantly carved of soapstone, these were superseded in the region by the so-called platform pipe, to which a long, flat stem was then attached. The upper part of the bowl, carved into bird or animal shape, was placed so as to face the smoker, looking, says Dr. Hall, for all the world like the hook at the end of a flat spear-thrower. It was the *atlatl* (as the Aztecs called it), together with the spear, which provided the weapon of note before bow and arrow arrived on the scene. A still older weapon, then, transformed into a pipe. Another symbol of armament prayerfully exchanged. All rather similar, Hall reminds us, to the twenty-one-gun salute: the cannons fired harmlessly skyward in honor, not with intent to kill.

Above all else, tobacco for the first Americans was the quintessential offering. One sprinkled tobacco reverently over the empty eye socket of the bear or into the ground from which one pulled a plant. One's pipe and the smoke therefrom were reverently dedicated to the sun. Tobacco juice might be poured on the ground to honor the ancients or down one's gullet—as did the Jivaro—to feed the power spirits residing there.

A peripatetic psychotrope

The Old World weed of many uses is *Cannabis,* or common hemp. A veritable supermarket of products, it is a source of fiber, oil, food, medicine, and delight. This last was, as Richard Schultes tells us, one of its early names in China. Known from Neolithic times (and possibly before that) in the Near and Far East, it appears to have been borne westward to Europe by various horse nomads of Central Asia. According to Herodotus, at least one group of these, the Scythians, burned hemp seeds in small tents and, creeping under the felt covers to inhale the smoke, howled for joy. Experts wonder how this could be so, since hemp seeds are nourishing food quite without euphoric or any other effect. Yet in the frozen tombs of the Altai, archaeologists have found the miniature tents as described by Herodotus, and a full complement of hemp seeds as well.

The variety of names attached to *Cannabis* describes manner of preparation as well as national classification. There is the familiar and most recent "marijuana" and also "choras," "ghanja," "bhang," "dagga," and "kef." How many devotees of the very strong "hash"—made by using only the flowering tops of female plants—know the origin of that name? It derives from Hashishin ibn-al Sabbah, the ambitious founder of a dissident Islamic cult. During the eleventh century the Old Man of the Mountain (as he came to be called) was able to attract some twelve thousand fanatically devoted followers. Not the least reason for their devotion was the nature of the leader's rewards. He provided *Cannabis* cocktails by which a soldier was transported to a seemingly magical garden populated by luscious and very biddable houris. Naturally, each man so longed to return there that he was willing to murder any religious rival the leader might care to specify. It is not only "hash" which derives from the leader's name but "assassin" as well. One wonders how many modern terrorist-assassins get their kicks from hash.

It may have been by way of Arabia (where the weed makes up for alcoholic prohibitions) that hemp made its way to Africa. It came later still to the New World. Rooting itself solidly in Mexico, it then made its way some sixty years ago into the United States—to what effect is well known.

Et cetera, worldwide

Psychotropic mushrooms of several genera are to be found in both hemispheres. In the New World—more especially in Mesoamerica—over twenty varieties are known and have been observed in native use. Described by the conquistadores, they can be related to even greater antiquity by the many mushroom effigies in archaeological contexts.

The Old World mushroom of magic is *Amanita muscaria,* the white dotted fly agaric, round and red. Its popularity among Siberian shamans (and among Russian colonists gone native) has been described. It was also used by Reindeer Lapps and perhaps in cults of the Scandinavian Bronze Age. Artifacts and petroglyphs indicate as much. Fly agaric is thought to be the divine and divinely intoxicating *soma* of the Aryan horsemen who invaded Persia and India around 1500 B.C. and left a record in the Vedas of their boisterous religious habits.

Familiar to herbalists of the two hemispheres are the powerful relatives of the deadly nightshade: mandrake, henbane, and thorn apple. The alkaloids of these plants are known to medicine as atropine and scopolamine, both psychotropic. And so potent are their properties that they can be absorbed through the skin. In this fact, says Michael Harner, lies the key to understanding the witchcraft craze of the late medieval, early Renaissance world. During those times, men and women were accused of assuming animal guise, of flying to horrid orgies called Sabbats and there succumbing to the Devil's fell designs. Under torture, there were confessions in plenty, followed by bonfires of the condemned. Scapegoats, present wisdom calls them, victims of a reality in which misfortune wore a human face and had a human source. The confessions, it is held, were prompted by pain and an all-too-lurid clerical imagination.

But the recorded confessions had something in common: references to a magic ointment which made one fly. Letters and documents of the period, written by physicians and other observant souls, indicate the ingredients: henbane, mandrake, nightshade, and thorn apple. Indeed, bodies of test subjects were rubbed with the ointment and observed in comatose condition while their spirits flew. Believing made it so. And for the belief, for the visions, for the "trips," thousands were done to death.

In the New World, thorn apple, or *Datura,* is known as jimsonweed in honor of the soldiers of Jamestown, who in 1676 cooked and ate it—with unanticipated but highly reportable results. It is also known as *toloache,* a Spanish corruption of the Aztec *toloatzin.* Tribes of central and southern California used the drug during initiation rituals staged by their religious cults. And throughout the area where the cult was most powerful, one finds, hidden in the hills, rock paintings of great strangeness and glowing color. There are, to be sure, the handprints and human stick figures common to many such ancient galleries. There are also abstract, snakelike designs, mazes, and geometrics. Could these represent the remembered visions of a spirit quest? Many specialists think so. Indeed, they are beginning to look with different eyes at all the rock art of ancient America. How much of it was drug-inspired?

Indeed, one might well ask much the same question about

other areas of art and vision. Travelers in Australia known of certain mushrooms, "blue meanies" and "gold tops," listed in the guidebooks as toxic. They are, in fact, hallucinogenic. Countless young visitors have tested the possibilities. Did Aboriginal people also test, finding in the process the myriad visions of the Dream Time?

In Hawaii there grows a native vine called "woodrose." Traveling cognoscenti claim for it hallucinatory properties similar to those of LSD. Did it, too, figure in native ritual? Ethnographic literature of other years will have to be recombed for hints of use. One sees, after all, what one is prepared to see. Observers then were not attuned, by habit or religious persuasion, to the psychotropic shortcut to ecstasy.

Since we have gone this far, what of the feral shamans, the were-beasts, of the Ice Age caves? How much of the ancient shaman's world view and, as a consequence, that of his flock had such beginnings? Probably more than we think, say modern ethnographers who have shared the psychotropic experiences of their hosts. Most report, among other phenomena, a sensation of body change. One may become animal—a great bird, a stalking cat, a snake. One may be in the body or outside of it. One flies high and far, encountering masked beings or masks without beings, encountering the powers and forces of the sacred made manifest. And all is vivid with color, mainly red—red splashed everywhere.

There is more, described by a young person who, like most of the older college set, has experimented with the milder mind-benders because it was the thing to do.

> We were driving [she said], my friend and I. In the road ahead of us was a fire engine, its red lights rotating madly, its sirens screaming. I knew it for what it was. I saw it for what it was. But I saw as well a huge lion roaring as it sprang. I told my friend and—what a surprise!—he had seen the same thing.

What they had seen was essence of fire engine. What they had seen was not illusion but metaphor. It is the power of metaphor that sends the Huichol Indians of mountainous western Mexico on a yearly pilgrimage three hundred miles away from home to find, as they put it, "their life."

The deer, the corn, and peyote

In truth, says Peter Furst, who was there, it is a journey home. Perhaps not a real home, but a sacred one. There lives peyote, a small, spineless cactus, powerfully hallucinogenic, which grows close to the ground. Its flesh is that of the Divine Deer, the Great Master Spirit who gives himself and must be honored. It is also a Maize Spirit, which itself must be reverenced. Here in this combined image are the two lives of the Huichol: the ancient hunting life and that of the farmer, settled and dependent on the land.

The pilgrims are led by a shaman, blessed by five previous journeys. Before departure, they must all fast and do without salt or chilis. They must be purified, confessing in public every sex partner since puberty. This is done not so much to banish sin, says Dr. Furst, as to discard their carnal natures. There are no recriminations. Good humor must prevail among the stay-at-homes as among the travelers, lest disaffection bring suffering on themselves and the crops.

They travel on modern highways in modern cars. No matter. The landscape is of the mind. When they find the refuge of Deer-Peyote, its flat green tops are surrounded by shot arrows. The archers weep and beg, "Do not punish us for killing you, for you have not really died. . . . We will feed you well . . . tobacco, water from Our Mothers, maize, and your favorite grasses. We have come from afar to greet you."

Consuming the peyote flesh as sacrament, they feel exaltation, see the gods, celebrate visions. Then they depart, tearfully promising the spirits of the place to return another year. Back at home, the visions may be recorded in yarn designs fixed with beeswax.

Much of the imagery and all of the sacredness of peyote was transmitted during the long years of eclipse and despair to Indian tribes north of the border, among whom it provided the central focus of a syncretistic native American religion.

An embarrassment of psychotropes

Peyote is not the only hallucinogenic cactus to be found natively south of the border. And the cacti are not the only hallucinogenic plants. We mentioned earlier the mushrooms. There are and were once used the morning-glory seeds, which contain substances very like lysergic acid diethylamide (better known as LSD). There are members of the pea family, such as mimosa and the mescal bean. And there are toads.

That some toads carry poison has long been known. Says Shakespeare's Duke in the Forest of Arden:

> Sweet are the uses of adversity,
> Which, like the toad, ugly and venomous,
> Wears yet a precious jewel in his head.

One wonders whether Shakespeare and his peers were aware of the mindbending properties of toad venom and whether the "precious jewel" might be a reference thereto. Certainly, some shamans in Mexico know of the visionary uses of toads. And the Amahuaca of eastern Peru have employed frogs in hunting magic (see Chapter 11). It is hard to set a time boundary. Michael Coe reports that in the Olmec site of San Lorenzo the most prevalent dietary remains (in addition to those of humans) are of the marine toad. Was the latter eaten there for food or for visions? Throughout Mesoamerica the Earth Mother was often visualized as a devouring toad. Hence the toad as altar and in art. Whether this relates as well to the ritual use of bufotenine, the toad alkaloid, is not yet known.

Bufotenine is to be found in the plant as well as the animal kingdom. Indeed, the New World is a veritable trove of intoxicants, stimulants, narcotics, and hallucinogens. A not surprising fact in view of its numerical superiority in terms of useful cultigens. In addition to the various cacti, mushrooms, trees, flowers, there are also vines. Those of the genus *Banisteriopsis* produce hallucinogens called variously in South America *ayahuasca, yage, natema caapi,* and, among the Cubeo, *mihi.*

Hallucinogenic materials in their raw states have been rendered into powder to be sniffed or smoked. They have been brewed into liquids which can be ingested, sniffed up the nose, or introduced by way of the enema. (Also not surprising in view of the fact that rubber is at home in the New World.) This particular method of incorporation is attested not only archaeologically but in ethnographic survivals. And why not? With an enema one gets quick drug action and no bitter taste. Recently, Michael Coe and Peter Furst located and described a Maya enema jar illustrated with figures of the Underworld Lords, elegant noses and pointed heads touching the floor as they crouch to receive from lady attendants their psychotropic injections.

Denominators, common and otherwise

How do mindbenders work? Do they all work in the same way?

Experiments in unraveling the stuff of visions have indeed indicated some general themes and forms. The most recent investigation in a scientific setting has been conducted by Dr. Ronald K. Siegel, a psychologist and pharmacologist on the faculty of the University of California, Los Angeles. Enlisting the help of both trained and untrained subjects, he administered cannabis (or its active component), psilocybin from the psychotropic mushroom, LSD, and mescaline. This, of course, by no means exhausts the list of mindbending substances.

Like participants in other, earlier experiments, Dr. Siegel's visionaries reported some constant forms: geometrics, spirals, tunnels, designs rather like the Hindu mandala. Colors: bright and mainly red. Sensations: distortions of the body, out of the body, floating, flying. Beings: mainly religious figures but also small animals and cartoons. Scenes: largely from memory.

As for possible root causes: stimulations of the central nervous system and in particular of the visual cortex. Possible, too, is a wrinkling of corneal tissues that change with rapid motions of the eyelid or a random firing of nerve cells in the eye itself. These serve to explain the form constants. The complex and dramatic imagery must be, says Dr. Siegel, rooted in memory, expectation, suggestion, desire. Alkaloids and, indeed, stress of other sorts may act to shut down the current of sensory input, remove the background radiation, so to speak, and allow enhanced focus on the "now." Alkaloids may heighten the effect of visionary "truth," but they do not make such truths or set them in motion. These are part of the reality system that a people hold in common. They stem as well from the fantasies and fears which lurk in each human heart and the degree of imagination that can be brought to bear. If the subject is a shaman of unusual ability and sensitivity, no wonder that his visions color the general expectation and world view.

Eloquent mushrooms

Hallucinations connote the visual, false or true. Dr. Siegel quickly discovered that untrained subjects were so overwhelmed by impressions that they had difficulty describing what they saw and could organize it not at all. For Mazatec Indians of southern Mexico, visions are of little importance. They take hallucinogenic mushrooms not to *see* but to *speak*—eloquently, with beauty, and with truth. "Flesh of the Gods" the mushroom is called now (as it was anciently in Mexico). It is also called "Prayer." It is called "Well-being," "Wisdom," "Holy One." For in its use the sacred things of Christianity and of the older native beliefs are comfortably merged.

"They are there with the Virgin, Our Mother, the Nativity," says the healer in family curing rites during which the mushrooms are eaten. The table on which they rest holds also tobacco and a cross. It is, in fact, an altar not so different from those reported by conquistadores long ago. So we learn from Henry Munn, who lived with the Mazatec and married the niece of a shaman.

When a person is ill, when things are going wrong, when there is suffering and fear, the shaman arranges family members around the table-shrine and all partake of the sacred food, the better to listen to the shaman's words:

> I am he who speaks with the dangerous mountains, says. I am he who speaks with the Father, says. I am he who speaks with the Mother, says. Where the terror of the day is, says. Where is the way of the dawn, the way of the day, says.

"Says" represents the teaching of the mushrooms and their imparted eloquence.

Have you ever tried to explain to another the nature of your fear and pain, of tangled motives, and cysts of guilt? And have you heard, as the words so earnestly intended dropped from your lips, the meanings twisted and turned false? They were not the wrong words, those words, but they were not the right ones, all the same. And you knew somehow that, given endless days and endless words, you would not frame a true message.

With words from the mushroom, the Mazatec shaman lays bare the inner secrets of the self and gives form to hope and to belief, form which binds and cures. Hallucinogens, in this instance, do not strike dumb but instead loose the tongue for healing eloquence. The shaman thus speaks for the patient, words which that one cannot himself discover or unlock. Words touch the sick heart and make of his scattered being one whole self again.

Culture and vision

Consider the Sharanahua healer who uses both vision and language to cure. Taking *ayahuasca,* he relives the dreams of the sick person, interpreting their forms to fit the patient's personal configurations of spirit and fantasy, adding always his own reassuring presence. He is there himself in the dream, and the threatening forces have been foiled, the dangers averted. It is, says Janet Siskind, who knows the Sharanahua, the deepest kind of communication, one in which powerful symbols are shared.

Consider the Jivaro man (or woman) of knowledge, gained through apprenticeship to a practicing shaman and the liquors of vision. With these the *tsentsak,* the many spirit helpers, can be seen and swallowed. In the shaman they are fed constantly with tobacco juice to maintain their alertness and their power to protect their master. From the shaman's stomach, they can be regurgitated as lights and animals and darts to kill or cure. But they are feckless beings, capricious and kill-hungry. To command them, the aspiring shaman must first control himself. The shaman-in-training abstains from sexual congress for months—a full year if he is really ambitious. Then he can murder with his darts or see with X-ray vision the inner workings of his patients and remove from them the darts of enemy shamans. To prove his powers, he casts his *tsentsak* at a tree, and if it splits, he is confirmed in confidence. The tree splits and falls, invisible to any but shamans. Never mind. It is the other world, the world of the *tsentsak*, that is the real one after all.

Consider Dedeheiwä, the renowned Yąnomamö shaman who was friend to Napoleon Chagnon. Daily he took hallucinogenic snuff, and with it was visited by the *Hekura* spirits who helped him cure or kill. Chagnon filmed him as, with other village shamans, he sought to cement a military alliance with a formerly hostile group by sending magical death to their joint enemy. We can see the shamans, beautifully painted to attract the *Hekura*, pose and preen, mimicking their several spirit helpers or the sufferings of the distant children now under attack. Through it all, Dedeheiwä directs the subordinate players, pulling and hauling them here and there in what becomes an elaborate drama. He manipulates their visions as well, subtly suggesting, guiding, cueing. And he can do this because he has control. He can use his own illusions, his own enhanced imagination, to his own ends.

It is not the vision that makes a shaman great. It is his power to command the vision as he commands his own rebellious flesh. It is his power to possess the being by which he is himself possessed. For everywhere the essence of a magical plant is anthropomorphized and named. Not everyone can meet such a challenge. Not everyone receives or can deal with visions, with the dominion of a force which seems outside oneself. But then, not everyone can master the body's own hungers or discipline its urge to sloth. Such discipline is a labor which drains the energy and excludes at least for a time all other claims of life.

Carlos Castaneda, the anthropology student who once apprenticed himself to a Yaqui sorcerer, suspended instruction because he could not master the visions and succumbed to fear, the "first enemy of a man of knowledge." Unbending determination is a heavy burden for one who seeks enlightenment all alone. That is perhaps why so many people court visions in groups, for groupness banishes fear, or at least keeps it at bay.

Says Irving Goldman, referring to the hallucinogenic *mihi*, "I spoke to no one, who pretended to enjoy it." The Cubeo drinker is seized with fear—of the violent, shadowy forms it summons, of the loss of self and home, of the trees in human shape, of the sense of rushing movement, of whiteness followed by the splash of red. But taking *mihi* is a necessary part, a necessary heightening of the sense of group. No wonder that the vines from which the drink is made are always represented as heavy, weighing down the man who carries them on his shoulder.

Writing of the core ritual of the Native American Church, the nightlong vigil during which congregants experience and share visions, David Aberle says:

> Taking peyote is hard: the taste is bitter, the nausea is unpleasant, the anxiety and depression overwhelming, and the night's ritual performance taxing . . . [but] the feeling of personal significance, the successful grappling with unpleasant feelings, the group experience, the vision as a communication with deity—all these are positive experiences for the cult member.

Mindbending, American style

It is the burden of fear and the exhausting struggle for control of which some sensitive young drug experimenters speak.

> I sat [said one] and looked at a blank wall. Soon it became transparent, a window to another world. I could see very clearly the houses and people, all strange in color and form. But I had to work and work hard to keep the picture clear. And once it was clear, I had to work still harder to keep it behind the wall.

It is not surprising to discover that the obligatory rebellion of the very young—the young of the suburban junior high school—is more and more often expressed in alcohol abuse. It may be worse in the long run for general health, but it is an easier hedonism and in many ways a safer one. Raiding Daddy's cellar is rather like running away from home while not leaving the parental block. Alcohol, after all, has a secure social place. In America, one winds down from one's labors with alcohol. ("Let's have a quick one before dinner.") One fortifies courage with alcohol. ("Give me a snort before I have to face the boss!") One calms shredded nerves. ("I needed that!") One celebrates. ("Let's pop a cork for your birthday!") Rewards effort. ("You deserve a good drunk.") Or socializes. ("Come over for drinks.")

Alcohol is secular. With several drinks, one may feel preternaturally pretty, witty, and wise. One's companions may express a similar self-beatitude. They may seem larger than life but are never scary. While tipsy, one need not expect to encounter archangels or monsters. And as for mood: exuberance, depression, quick anger, sodden sleep. No ecstasies. But also no horrors. One may ruin one's liver, lose one's job, wreck the car. One will seldom see "the little man who wasn't there." And if one does, he will not require genuflection.

Again, it is no surprise to discover that many youngsters of the sixties, once heavily into drugs, have kicked the habit only through membership in religious cults of various origins, cults which not only forbid psychotropes but alcohol as well. Forbidden, indeed, are indulgences of any kind, including those of thought. And the refugees from unlimited choice flock to the cults in great numbers, having discovered that utter freedom of flesh and mind can also be tyranny for those untrained in structuring a life.

One such religious group, homegrown and nearly as American as apple pie, is the Farm, which Kate Wenner described for *The New York Times.* Located in rural Tennessee, its members had traveled there from San Francisco in search of complete freedom in the communal life. Within seven years, the commune had become something close to a monastery both in feeling and in IRS classification. Banned were free love, free thinking, personal vanity. Required were hard work, reverence for God, obedience to the leader, allegiance to spouse and children. A vegetarian diet was in. Liquor, stimulants, and cigarettes were out. So were all drugs but one—marijuana, which they now call the "sacred weed" and use as a sacrament.

Given freedom, we seek constraint. Given ease, we opt for effort. That is the human part of human nature: to earn power through command. And if we cannot command ourselves, we locate a person who will do it for us, or a structure or a group. For meaning lies in constraint. Creativity flourishes within boundaries. Gains depend on pains. Stoicism and hedonism have ever been at war—in individuals as in groups—with stoicism the ultimate marker of new beginnings.

> For what is a man profited [asked Jesus], if he shall gain the whole world and lose his own soul? Or what shall a man give in exchange for his soul?

EPILOGUE
Body and Soul

Or what shall a man give in exchange for his soul?

Immortality of the soul was thus the message of Jesus. And the imperishable aspect of person was his concern, the future life ever in jeopardy to sin.

The word *soul* has carried still other meanings and prompted many questions, too. Is there a soul separable from the body? Is there a shadowy mirror image that survives, recognizable and intact? If so, where does it go at death? Or does it go anywhere at all?

These have been matters of hot debate in the West ever since the end of religious persecution made such questions tolerable. The East has long cherished—even during its phases of atheism—a more concerted vision of soul. It is a vision of eternal return, an endless round of rebirth. In a celestial economy which allows neither waste nor loss, what could make better sense than spiritual recycling? Another reason for the enduring appeal of this view of soul may lie in its inherent power to explain evil in the world, terrible and seemingly unmerited.

Separable or no, we may be sure of one thing. When the soul falters, the body dies. Deaths of fright, of fear, of breaking down and giving up, are as well known to Western medicine as they are among folk who still believe in sorcery and the punishment which afflicts the profaner of sacred things.

> The Lord let the house of a brute to the soul of a man,
> And the man said, "Am I your debtor?"
> And the Lord—"Not yet: but make it as clean as you can,
> And then I will let you a better."

Is soul, then, the essence, the uniqueness of species, an entity fully formed in its beginning and, through all the transformations of body, forever the same? Alfred, Lord Tennyson seems to have pursued this notion in his "Lines by an Evolutionist." And some of us pursue it still when we speak of "soul food," "soul music," and "soul," intending thereby to convey a sense of special ethnic identity.

> Out of the night that covers me,
> Black as the Pit from pole to pole,
> I thank whatever gods may be
> For my unconquerable soul.

Courage seems to have been the meaning of "soul" for William Ernest Henley. Is soul, then, equivalent to will? To intelligence? Personality? Psyche? Is it merely that invisible fabric invisibly woven from the threads of inherent possibility, social givens, and unique experience?

The so-called primitives of the world often have a better handle on the concept of soul, both in its parts and in its totality. For the Ojibwa of Canada, the soul is dual in nature. One part lives near the heart, escapes in dreams, drunkenness, and insanity, and, if not recovered, brings death to the body. Yet another part, the elusive Shadow, lives in the head and serves as spiritual "eye." It plays cool intelligence to the soul's passion and will. These will survive intact in the eternal West. But intelligent Shadow, lingering for a time near the body's grave, fades as memory fades and is lost.

The Dogon soul is also dual so as to celebrate the ambivalences of sex. Each person, say the Dogon, combines male and female souls, though only one is normally in residence. The other hovers in the sky or floats in water, establishing from those primordial centers vital links with the body. No wonder the Dogon insist on genital operations to confirm the body in its sexual identity. No wonder the unmarried man begins early to seek out the mate who will be, body and soul, his living completion.

Life, image, and immortality are alike encompassed by the Eskimo concept of Breath Soul. Unique personality, on the other hand, intelligence, skill, magic, strength—all these are vested in a Name Soul, which has its immortality in the successive living bodies to whom name is transmitted. And the more names one can acquire, the more secure one's hold on life. All animals but the dog have only Breath Souls, and each species has its own "person"—a projected essence, a "Master," often human in form. But humankind themselves lack a species master. So do dogs, who also share with humankind the Name Soul, the power of unique, special identity denied to other animals.

The Kwakiutl deny Name Soul as well to human beings of commoner rank. To be sure, these have their own personal essence, the soul which will depart as a screech owl to the otherworld. Only to an aristocrat is given the Soul of Name and the Soul of Form, each of which is eternal, specific, to be activated by its bearer, life after life. The animal's skin, the tree's bark, the carved heirloom dish—these are also Form Souls, but they are generic and inert. Mask, however, is a Form Soul, individual and unique. It represents a once and future being and is itself susceptible to excitement, moved by volition. Yet another kind of Form Soul is the house which shelters a ranked lineage, together with the names and masks revived in each generation forever and ever.

The Fon of Dahomey, the old West African kingdom recently redubbed Benin, are even more separatist in the matter of soul. The Clan lives in each individual through a shadowy ancestor who serves as mentor and guide. With him is his protégé's own share of the great World Soul of God. There is also a personal soul, locus of unique attributes and destiny. There is Shadow, which is indestructible. At death, World Soul flies back to God. Personality Soul is extinguished. But Shadow then becomes Ancestor, to be born anew within the Clan. It is the family which is forever.

Australian Aborigines separate nothing and quite happily merge the concepts of ghost, name, shadow (the real projection on the ground), body, the Dream Time, the place from which one's spirit came, the totem beings with which one affiliates socially or personally. All these elements animate the person and, in essence, go with him when he dies, returning to the spirit center from which he will eventually achieve rebirth. But one must contrive to die in one's own land, in the bosom of one's family (however that is constituted), else that ultimate unity is lost forever, annihilated.

With this all-purpose oneness, as with the various particularities of soul, we cannot relate. Yet our culture no longer offers a unitary design. And we are hard put to find one for ourselves.

Even less susceptible to a single definition is the matter of humanness. "It's human nature to ______." (Be curious, struggle, desire, look out for Number One.) Fill the blank with your favorite prescription. Any effort to mitigate behavior is usually prefaced with the statement "After all, I'm only human!" But what exactly is that? The more we know, the less certain we become.

And so it is that each year brings its rash of images, a new set of books defining what it means to be human. We are invited to view ourselves as killer apes, as naked apes, as territorial imperialists, as hunters unmodified and unreconstructed.

Some specialists are interested in discovering how far up our family tree the tyranny of instinct can be traced. Others follow the glimmerings of mind as far down the trunk as they dare. In either direction, we lose something of the distinctions that our primitive forebears strove to draw in paint, in mutilation, in fasting and prayer. "If we intersect the rest of creation," they seemed to feel, "let it be at a time of our choosing and in the service of metaphor. For it is *we* who are special, who are different, and who know it!" Today we are not so sure.

Christianity, with its historical (though not evolutionary) emphasis, favored the distinction. It echoed the "lesser angel" theme introduced by the Old Testament Psalmist, yet retained the mark of the beast, of original sin, which itself offers a way of explaining evil in the world.

When humankind developed the machine and moved with it into an Industrial Revolution, the engineering image came to the fore. Everything about the person, it was felt, could be patched, repaired, retooled, tinkered to death. Even the ailing mind (does a machine have a soul?) was thought to respond favorably to a good tune-up. Indeed, neurologists have for long envisioned the brain itself as an enormously sophisticated machine. With the advent of cybernetics, it became a computer with the world's ultimate in miniaturized circuitry. In the most recent symposia of the experts, the machine image has been replaced by one of community, a more homely concept, surely, with anticipated unpredictability built in.

Psychologists have tended to emphasize the irrational component in human nature, focusing on the titanic battle of Id, Ego, and Superego (dramatis personae the primitive might well encompass in a divisible soul). There in humanity's subbasement lives Id, driven by ruthless primal urges, forever breaking out to push or possess the vulnerable ego-self. This harassed being is subject as well to the ruthless demands of Superego, censorious and autocratic, a composite of society's unrealistic ideals and rules.

Cultural Materialists favor a view of human nature that is doggedly rational in spite of itself. The sum total of blind individual efforts to adjust will, they say, always favor logical outcomes. One has only to look hard to find the underlying reasons. They always make sense.

Biologists, many of them, also lean toward rationality but like to vest it in the genes. For a time it was popular to see innate aggression—as evident in human behavior as in that of the guppy—as the key to survival (and possibly ultimate destruction, as the case may be). Now innate altruism is endorsed. Selection, it is maintained, invariably favors the relatives of individuals programmed to put the good of the family ahead of their own good. Even if that means altruistic suicide.

We advance, to be sure, by way of our models, our metaphors. Each successful model begins by being an explanation and ends as a World View: a new approach to total reality. But models impose their own tyranny. Each in its way not only colors our perception of past and future but conditions today's behavior as well. We tend to dream as we are expected to dream, and feel as we are expected to feel, always to suit the model rather than the other way around. And come no closer to primordium.

The untidy truth seems to lie somewhere in the fact that we have evolved a mind and soul which dominate body in spite of its fail-safe mechanisms, its systems of compensation and balance, its gut determination to survive. Mind over matter.

But brain is also matter, and it is brain which powers mind. Let us then look at brain. At once we see that structures "new" in humankind are those which foster idea correlation and foster speech. Now, speech is image made audible. And image captures what *is* and directs willy-nilly the course of what will be.

Image: this is what our species is all about. Never mind what adaptive behavior obtains for flatworms, titmice, and wasps. "In the beginning," said Saint John, "was the Word." It was certainly our beginning. And it is the unique "soul" of the species.

If you prefer the engineering model, then think of us as image machines, cranking out meaning the way a gun shoots bullets. We cannot help it. We cannot stop. We never shall.

Any historian knows it is the image that greases the wheels of action. Knows that a people's *perceived* oppression is more powerful than the actual existence of gibbet and chain. Knows, too, that the social bond is more concept than compact, a thing dependent on the sanctifying image—once "God," then "King," then "People," and now "Me."

The expert in public relations uses image to sell popes and presidents as well as beer, movies, and tourist traps. It is *what could be,* not what *is,* that counts. Believing makes it so.

Even the economy (unrelentingly practical though it is) has its own elements of fantasy. The Great Depression was certainly no figment of the imagination. But recovery for a long time was all in the collective perception. Happy days are here again! Believing made it so. The image machine at work.

As a famous scientist once said, "We are the universe made conscious of itself." It is a growing acceptance of this image, a growing sense of loneliness in immensity that lies behind the flying saucer craze. We do not want to bear all alone the crushing responsibility for comprehension. Let there be other sentient beings, we implore. No matter how fearsome they are, how patronizing, how importunate, let there be others!

"What is the answer?" asked Gertrude Stein on her deathbed. And receiving no reply from the mourners clustered about, she sighed and said, "Then what is the question?"

The question, like beauty, is its own excuse for being. And it is ours. Out of a brain of a particular kind grew awareness capable of transforming its own vehicle into an instrument of metaphor. So that blood became something greater than the fluid of life, and the phallus something more than an organ of generation at the behest of every passing hormone. Out of a brain of a particular kind grew awareness capable of projecting on the face of the unknown images that started with its vehicle and can now encompass infinity.

However puny we are as individuals, however ignoble, however mean, petty, grasping, cruel, there is yet that collective power to create image and to transcend. It is in its way a kind of valor. It is the beginning and the end of being human.

BIBLIOGRAPHY

PROLOGUE: *Be All, End All*

Douglas, Mary, *Purity and Danger.* New York: Praeger, 1966.

———, *Natural Symbols: Explorations in Cosmology.* New York: Random House, 1973.

Turner, Victor W., *The Forest of Symbols.* Ithaca and London: Cornell University Press, 1967.

1. THE GRAPHIC BODY: *Paint Me Human*

Arens, William, "The Great Football Ritual." *Natural History Magazine,* October 1975.

Bordes, François, *A Tale of Two Caves.* New York: Harper & Row, 1972.

Chagnon, Napoleon A., *Yąnomamö: The Fierce People.* New York: Holt, Rinehart and Winston, 1977.

Cormack, Margaret Lawson, *The Hindu Woman.* New York: Bureau of Publications, Teachers College, Columbia University, 1953.

Darwin, Charles, *The Voyage of HMS Beagle.* New York: Heritage Press, 1957.

de Lumley, Henry, "A Paleolithic Campsite at Nice." *Scientific American,* May 1966.

Densmore, Frances, *Chippewa Customs.* Bureau of American Ethnology Bulletin no. 86. Washington: Government Printing Office, 1929.

Dubois, J. A., *Hindu Manners, Customs, and Ceremonies.* London: Oxford University Press, 1906.

Edwards, Stephen W., "Nonutilitarian Activities in the Lower Paleolithic." *Current Anthropology,* March 1978.

Elkin, A. P., *The Australian Aborigines.* New York: American Museum of Natural History, Doubleday, 1964.

Goldman, Irving, *The Cubeo.* Urbana: University of Illinois Press, Illinois Studies in Anthropology no. 2, 1963.

Gusinde, Martin, *The Yahgan: The Life and Thought of the Water Nomads of Cape Horn.* Mödling Gei Wein, Anthropos-Bibliothek, 1937, 20.

Hart, C.W.M., and Arnold R. Pilling, *The Tiwi of North Australia.* New York: Holt, Rinehart and Winston, 1960.

Heider, Karl G., *The Dugum Dani.* Viking Fund Publications in Anthropology no. 49. New York: Wenner Gren Foundation, 1970.

Holmberg, Allan R., *Nomads of the Long Bow.* Garden City, N.Y.: Natural History Press, 1969.

Howell, F. Clark, "Observations on the Earlier Phases of the European Lower Paleolithic." *American Anthropologist,* April 1966.

Ingham, John M., "Are the Siriono Raw or Cooked?" *American Anthropologist,* October 1971.

Karsten, Rafael, *The Headhunters of the Western Amazonas.* Helsinki: Centraltryckeriet, 1935.

Lahren, Larry, and Robson Bonnichen, "Bone Foreshafts from a Clovis Burial in Southwestern Montana." *Science,* Oct. 11, 1974.

Leakey, Mary D., *Olduvai Gorge.* Cambridge, England: Cambridge University Press, 1971.

Marshak, Alexander, "Some Implications of the Paleolithic Symbolic Evidence for the Origin of Language." *Current Anthropology,* June 1976.

Meggitt, M. J., *Desert People.* Chicago: University of Chicago Press, 1971.

Musil, Alois, *The Manners and Customs of the Rwala Bedouins.* Vol. 14. New York: American Geographical Society, 1928.

O'Malley, L.S.S., *Popular Hinduism.* Cambridge, England: Cambridge University Press, 1935.

———, ed., *Modern India and the West.* London: Oxford University Press, 1941.

Radcliffe-Brown, A. R., *The Andaman Islanders.* Glencoe, Ill.: The Free Press, 1964.

Siskind, Janet, "Special Hunt of the Sharanahua." *Natural History Magazine,* October 1974.

Strathern, Andrew, and Marilyn Strathern, *Self Decoration in Mt. Hagen.* Toronto: University of Toronto Press, 1971.

Turner, Terence S., "Tchikrin: A Central Brazilian Tribe and Its Symbolic Language of Bodily Adornment." *Natural History,* October 1969.

Turner, Victor W., "Color Classification." In *The Forest of Symbols.*

Wreschner, Ernst E., "The Red Hunters: Further Thoughts on the Evolution of Speech." *Current Anthropology,* December 1976.

2. THE PLASTIC BODY: *Improving on Nature*

Adamson, Joy, *The Peoples of Kenya.* New York: Harcourt, Brace, 1967.

Adriani, N., and Albert C. Kruyt, *The Bare'e-Speaking Toradja of Central Celebes.* 3 vols. Amsterdam: Noord-Hollandsche Uitgevers Maatschappij, 1950.

Akiga, *Akiga's Story.* Translated by Rupert East. London: International African Institute, 1939.

Alkema, B., and T. J. Bezemer, *Concise Handbook of the Ethnology of the Netherlands East Indies.* Haarlem: H. D. Tjeenk Willink 'en Zoon, 1927.

Batchelor, John, *Ainu Life and Lore.* Tokyo: Kyobunkwan, [1927?].

Best, Elsdon, *The Maori.* Polynesian Society Memoir no. 5. 2 vols. Wellington, New Zealand: Board of Maori Ethnological Research, 1924.

Bohannon, Paul, and Laura Bohannon, *Three Source Notebooks in Tiv Ethnography.* New Haven, Conn.: Human Relations Area Files, 1958.

Buck, Peter H. (Te Rangi Hiroa), *The Coming of the Maori.* Wellington, New Zealand: Whitcombe and Tombs, 1949.

Dorsey, George Amos, *The Arapaho Sun Dance.* Millwood, N.Y.: Kraus Reprint, 1973.

——— and A. L. Kroeber, *Traditions of the Arapaho.* Chicago: Field Columbian Museum, 1903.

Downs, Richard Erskine, *The Religion of the Bare'e-Speaking Toradja of Central Celebes.* The Hague: Uitgeverij Excelcior. 1956.

Drucker, Philip, *Indians of the Northwest Coast.* Garden City, N.Y.: Natural History Press, 1963.

Dubois, Cora, *The People of Alor.* Minneapolis: University of Minnesota Press, 1944.

Gilliard, E. Thomas, "New Guinea's Rare Birds and Stone Age Men." *National Geographic,* April 1953.

Goldman, Irving, *Ancient Polynesian Society.* Chicago: University of Chicago Press, 1970.

Griaule, Marcel, *Conversations with Ogotemmeli.* London: Oxford University Press, 1965.

Henry, Jules, *Jungle People.* New York: Random House, 1964.

Herodotus of Halicarnassus, *The Histories.* 2 vols. Translated by Harry Carter. New York: Heritage Press, 1958.

Hollis, Alfred Claud, *The Masai: Their Language and Folklore.* Oxford: Clarendon Press, 1905.

Huard, Pierre, "Le Noircissement des Dentes en Asie Orientale et en Indochine." *France-Asie* 3, 1948.

Leroi-Gourhan, André, "French Somaliland." In *Ethnologie de l'Union Française*. Paris: Presses Universitaires de France, 1953.

Levy, Howard S., *Chinese Footbinding*. New York: Walton Rawls, 1966.

Lhote, Henri, "The Fertile Sahara." In *Vanished Civilizations*, edited by Edward Bacon. New York: McGraw-Hill, 1963.

Maxwell, Nicole, *Witchdoctor's Apprentice*. New York: Collier, 1975.

Meggitt, M. J., *Desert People*.

Muensterberger, Warner, "Roots of Primitive Art." In *Anthropology and Art*, edited by C. Otten. Garden City, N.Y.: Natural History Press, 1971.

Musil, Alois, *The Manners and Customs of the Rwala Bedouins*.

Paulitschke, Philipp, *Contributions to the Ethnography and Anthropology of the Somali, Galla, and Harari*. Leipzig: Eduard Baldamus, 1888.

Philby, Harry St. John Bridger, *Sa'udi Arabia*. London: Blau, 1950.

Pruitt, Ida, *A Daughter of Han*. Stanford, Calif.: Stanford University Press, 1967.

Puccioni, Nello, *Anthropology and Ethnography of the Peoples of Somalia*. 3 vols. Bologna: Nicola Zanichelli, 1936.

Richie, Donald, "The Japanese Art of Tattooing." *Natural History*, December 1973.

Seeger, Anthony, "The Meaning of Body Ornaments: A Suya Example." *Ethnology*, July 1975.

Shultz, Harald, "Brazil's Big Lipped Indians." *National Geographic*, January 1962.

Singer, Isidore, ed., *The Jewish Encyclopedia*. New York: KTAV Publishing House, Inc., 1901.

Steward, T. D., *The People of America*. New York: Scribners, 1973.

Stoloft, Charles I., "The Fashionable Tooth." *Natural History*, February 1972.

Strathern, Andrew, and Marilyn Strathern, *Self Decoration in Mt. Hagen*.

Thomas, Elizabeth Marshall, *The Warrior Herdsmen*. New York: Knopf, 1965.

Tozzer, Alfred M., ed. and trans., *Landa's Relacioñ de las Cosas de Yucatân*. Papers of the Peabody Museum of American Archaeology and Ethnology. Vol. 18. Cambridge, Mass.: Harvard University Press, 1941.

Turner, Terence, "Tchikrin: A Central Brazilian Tribe and Its Symbolic Language of Bodily Adornment."

3. WALL-TO-WALL BODY: *The Image Transferred*

Bandi, Hans-Georg, et al., *The Art of the Stone Age.* New York: Crown, 1961.

Berndt, Ronald M., and Catherine H. Berndt. *The World of the First Australians.* Chicago: University of Chicago Press, 1964.

———, ed., *Australian Aboriginal Art.* New York: Macmillan, 1964.

Bogoras-Tan, Vladimir Germanovich, *The Chukchee.* 3 vols. New York: E. J. Brill, 1904–1909.

Bordes, François, *The Old Stone Age.* New York: McGraw-Hill, 1968.

Breuil, Abbé H., *Four Hundred Centuries of Cave Art.* Translated by Mary E. Boyle. Montignac, Dordogne: Centre d'Etudes et de Documentation Préhistorique, 1952.

Broderick, Alan Houghton, *Father of Prehistory.* New York: Morrow, 1963.

Elkin, A. P., *The Australian Aborigines.*

Fairservis, Walter, *The Threshold of Prehistory.* New York: Scribners, 1975.

Feit, Harvey A., "Twilight of the Cree Hunting Nation." *Natural History,* August–September 1973.

Gould, Richard A., *Yiwarra.* New York: Scribners, 1969.

Graziosi, Paolo, *Paleolithic Art.* New York: McGraw-Hill, 1960.

Jelinek, J., *The Pictorial Encyclopedia of the Evolution of Man.* London: Hamlyn, 1975.

Jenness, Diamond, *The Indians of Canada.* National Museum of Canada Bulletin no. 65, Anthropological Series 15, 1955.

Klein, Richard G., "Ice Age Hunters of the Ukraine." *Scientific American,* June 1974.

Laming, Annette, *Lascaux.* Translated by Eleanor F. Armstrong. Baltimore: Penguin, 1959.

Laming-Emperaire, Annette, *La Significance de l'Art Rupestre Paléolithique.* Paris: Editions A. et J. Picard et Cie., 1962.

Leroi-Gourhan, André, *Treasures of Prehistoric Art.* Translated by Norbert Guterman. New York: Abrams, 1967.

Levine, Morton, "Prehistoric Art and Ideology." *American Anthropologist,* December 1957.

Lhote, Henri, "The Fertile Sahara." In *Vanished Civilizations*, edited by Edward Bacon. New York: McGraw-Hill, 1963.

———, *In Search of the Tassili Frescoes.* 2d ed. London: Humanities Press, 1973.

Maringer, Johannes, *The Gods of Prehistoric Man.* New York: Knopf, 1960.

Marshak, Alexander, *The Roots of Civilization.* New York: McGraw-Hill, 1972.

———, "Cognitive Aspects of Upper Paleolithic Engraving." *Current Anthropology*, June–October 1972.

———, "Upper Paleolithic Notation and Symbol." *Science*, November 24, 1972.

———, "Exploring the Mind of Ice Age Man." *National Geographic*, January 1975.

———, "Some Implications of the Paleolithic Symbolic Evidence for the Origin of Language." *Current Anthropology*, June 1976.

———, "The Message in the Markings." *Horizon*, Autumn 1976.

———, "The Art and Symbols of Ice Age Man." *Human Nature*, September 1978.

Meggitt, M. J., *Desert People.*

Mellaart, James, *Çatal Hüyük.* New York: McGraw-Hill, 1967.

———, *The Neolithic of the Near East.* New York: Scribners, 1975.

Mori, Fabrizio, "The Earliest Saharan Rock-Engravings." *Antiquity*, June 1974.

Mulvaney, D. J., *The Prehistory of Australia.* New York: Praeger, 1969.

———, "Prehistory Down Under." *Natural History*, April 1970.

Pales, Léon, and Marie Tassin de Saint Péreuse, "Humains Superposés de La Marche." In *Prêhistoire, Problèmes et Tendences.* Paris: Éditions du Centre National de la Recherche Scientifique, 1968.

Pericot-García, Luis, and Eduardo Ripoll Perello, eds., *Prehistoric Art of the Western Mediterranean and the Sahara.* Viking Fund Publications in Anthropology no. 39. New York: Wenner-Gren Foundation, 1964.

Reed, Robert C., "An Interpretation of Some Anthropomorphic Representations from the Upper Paleolithic." *Current Anthropology*, March 1976.

Titiev, Mischa, *Old Oraibi.* Papers of the Peabody Museum of American Archaeology and Ethnology, vol. 22, no. 1, 1944.

Ucko, Peter J., and Andree Rosenfeld, *Paleolithic Cave Art.* New York: McGraw-Hill, 1967.

4. SAVING FACE: *Masks and Meanings*

Caldwell, Joseph R., and Robert L. Hall, eds., *Hopewellian Studies.* Springfield, Ill.: Illinois State Museum Papers, vol. 12, 1970.

Campbell, Joseph, *The Masks of God: Primitive Mythology.* New York: Viking Press, 1959.

Catlin, George, *O-kee-pa: A Religious Ceremony and Other Customs of the Mandan.* New Haven, Conn.: Yale University Press, 1967.

Coe, Michael D., *America's First Civilization.* New York: American Heritage Publishing Co., 1968.

———, "The Shadow of the Olmecs." *Horizon,* Autumn 1971.

———, *The Maya Scribe and His World.* New York: The Grolier Club, 1973.

———, *Lords of the Underworld.* Princeton, N.J.: Princeton University Press, 1978.

Covarrubias, Miguel, *The Eagle, the Jaguar, and the Serpent.* New York: Knopf, 1954.

Cunningham, Ann Marie, "Masked Messages." *Natural History,* March 1977.

D'Azevedo, Warren L., "Gola Mask Makers." In *Primitive Art and Society,* edited by Anthony Forge. London: Oxford University Press, 1968.

Dieterlen, Germaine, *An Essay on the Religion of the Bambara.* Paris: Presses Universitaires de France, 1951.

Feast, The. A film about the Yąnomamö. National Audiovisual Center.

Fenton, William Nelson, *Masked Medicine Societies of the Iroquois.* Smithsonian Institution. Annual Report of the Board of Regents, 95, 1938–1939.

Furst, Peter T., "The Were-Jaguar Motif in the Light of Ethnographic Reality." In *Dumbarton Oaks Conference on the Olmec,* edited by Elizabeth Benton. Washington: Dumbarton Oaks Research Library and Collection, 1968.

Gay, Carlo T. E., "Rock Carvings at Chalcatzingo." *Natural History,* August–September 1966.

Gimbutas, Marija, "The Mask in Old Europe: 6500–3500 B.C." *Archaeology,* October 1974.

———. "Varna: A Sensationally Rich Cemetery of the Karanova Civilization, about 4500 B.C." *Expedition*, Summer 1977.

Goldman, Irving, *The Mouth of Heaven: An Introduction to Kwakiutl Religious Thought*. New York: Wiley, 1975.

Grant, Campbell, *Rock Art of the American Indian*. New York: Crowell, 1967.

Griaule, Marcel, *Conversations with Ogotemmeli.*

Griffin, Gillett G., Personal communication.

Grove, David C., "The Olmec Paintings of Oxtotitlán Cave, Guerrero, Mexico." In *Studies in Precolumbian Art and Archaeology* no. 6. Washington, D.C.: Dumbarton Oaks, 1970.

Gusinde, Martin, *The Yahgan.*

Imperato, Pascal James, "Last Dances of the Bambara." *Natural History*, April 1975.

Jenness, Diamond, *The Life of the Copper Eskimo*. Report of the Canadian Arctic Expedition, 1913–1918. Vol. 12.

———, *Myths and Traditions from North Alaska*. Report of the Canadian Arctic Expedition, 1913–1918. Vol. 13.

———, *People of the Twilight*. Chicago: University of Chicago Press, 1959.

Joralemon, Peter David, "A Study of Olmec Iconography." In *Studies in Precolumbian Art and Archaeology* no. 7. Washington, D.C.: Dumbarton Oaks, 1971.

Kraft, Herbert C., "The Rosenkrans Site, an Adena-Related Mortuary Complex in the Upper Delaware Valley, New Jersey." *Archaeology of Eastern North America*, Winter 1976.

Lindskog, Birger, "African Leopard Men." *Studia Ethnographica Upsaliensia* 7, 1954.

Little, Kenneth, "The Political Function of the Poro." In *Africa and Change*, edited by Colin Turnbull. New York: Knopf, 1973.

Magical Death. A film about the Yąnomamö. Napoleon Chagnon. University Park, Pa.: Pennsylvania State University, 1973.

Murphy, Robert F., "Social Distance and the Veil." *American Anthropologist*, December 1964.

Otten, Charlotte, ed., *Anthropology and Art*. Garden City, N.Y.: Natural History Press, 1971.

Palau Marti, Montserrat, *The Dogon*. International African Institute. Paris: Presses Universitaires de France, 1957.

Paul, Anne, "History on a Maya Vase." *Archaeology*, April 1976.

Sahagún, Bernardino de, *Florentine Codex: General History of the Things of New Spain.* Translated by C. E. Dibble and Arthur J. O. Anderson. Monograph of the School of American Research and the Museum of New Mexico no. 14, 1961.

Sieber, Roy, "Masks as Agents of Social Control." *African Studies Bulletin,* vol. 5, no. 11, May 1962.

Steward, Julian H., ed., *Handbook of South American Indians.* Smithsonian Institution. Bureau of American Ethnology. Washington: Government Printing Office, 1946.

Thompson, Robert Farris, "The Sign of the Divine King." *African Arts,* Spring 1970.

Titiev, Mischa, *Old Oraibi.*

Tozzer, Alfred M., *Landa's Relación de las Cosas de Yucatán.*

Waite, Deborah, *"Kwakiutl Transformation Masks."* In *The Many Faces of Primitive Art,* edited by Douglas Fraser. Englewood Cliffs, N.J.: Prentice-Hall, 1966.

Wauchope, Robert, ed., *Handbook of Middle American Indians.* Austin: University of Texas Press, 1965.

5. FLUIDS AND EMANATIONS: *Wetness as Metaphor*

Abraham, Roy Clive, *The Tiv People.* Lagos: Government Printer, 1933.

Adriani, N., and Albert C. Kruyt, *The Bar'e-Speaking Toradja.*

Berndt, Ronald M., *Excess and Restraint.* Chicago: University of Chicago Press, 1962.

Best, Elsdon, *The Maori.*

Coe, Michael D., *The Jaguar's Children: Pre-Classic Central Mexico.* New York: The Museum of Primitive Art, New York Graphic Society, 1965.

Driver, Harold E., *Indians of North America.* 2d ed. Chicago: University of Chicago Press, 1969.

Dubois, J. A., *Hindu Manners, Customs, and Ceremonies.* London: Oxford University Press, 1906.

Durant, Will, *Our Oriental Heritage.* New York: Simon and Schuster, 1942.

Elkin, A. P., *The Australian Aborigines.*

Griaule, Marcel, *Conversations with Ogotemmeli.*

Hart, C.W.M., and Arnold R. Pilling, *The Tiwi of North Australia.*

Heider, Karl G., *The Dugum Dani.*

Heizer, R. F., and M. A. Whipple, *The California Indians.* Berkeley: University of California Press, 1970.

Herodotus of Halicarnassus, *The Histories.*

Hudson, Charles, "Vomiting for Purity: Ritual Emesis in the Aboriginal Southeast United States." In *Symbols and Society,* edited by Carole E. Hall. Athens: University of Georgia Press, 1975.

Hutton, John Henry, *Caste in India.* Bombay: Oxford University Press, 1951.

Meggitt, M. J., *Desert People.*

Miner, Horace, "Body Ritual Among the Nacirema." *American Anthropologist,* June 1956.

Newman, Philip L., *Knowing the Gururumba.* New York: Holt, Rinehart and Winston, 1965.

Oswalt, Wendell H., *This Land Was Theirs.* New York: Wiley, 1973.

Palau Marti, Montserrat, *The Dogon.*

Paulme, Denise, *Social Organization of the Dogon.* Éditions Domat-Montchrestien. Paris: F. Loviton et Cie., 1940.

Radcliffe-Brown, A. R., *The Andaman Islanders.*

Sahagún, Bernardino de, *Florentine Codex.*

Stevenson, Margaret, *The Rites of the Twice Born.* London: Oxford University Press, 1920.

Tozzer, Alfred M., *Landa's Relación de las Cosas de Yucatán.*

Wagner, Roy, *The Curse of Souw.* Chicago: University of Chicago Press, 1967.

6. PARTS AND PIECES: *The Anatomy of Meaning*

Arbman, Holger, "Bronze Age Seen in Granite." *Natural History,* November 1964.

Arens, William, "The Great Football Ritual."

Barton, R. F., *The Half-Way Sun: Life Among the Headhunters of the Philippines.* New York: Brewer, Warren and Putnam, Inc., 1930.

Berndt, Ronald, *Excess and Restraint.*

Bohannon, Paul, and Laura Bohannon, *Three Sourcebooks in Tiv Ethnography.*

Breasted, James Henry, *A History of Egypt.* New York: Scribners, 1909.

———, *Development of Religion and Thought in Ancient Egypt.* New York: Harper, 1959.

Burl, Aubrey, *Stone Circles of the British Isles.* New Haven, Conn.: Yale University Press, 1976.

Campbell, Bernard G., *Human Evolution.* 2d ed. Chicago: Aldine, 1974.

———, *Humankind Emerging.* Boston: Little, Brown, 1976.

Driver, Harold E., *Indians of North America.*

Durant, Will, *Our Oriental Heritage.*

Evans-Pritchard, E. E., *Witchcraft, Oracles and Magic Among the Azande.* Oxford: Clarendon Press, 1937.

Goldman, Irving, "The Ifugao of the Philippine Islands." In *Cooperation and Competition,* edited by Margaret Mead. Boston: Beacon Press, 1961.

Gould, Richard A., *Yiwarra.*

Granet, Marcel, *Chinese Civilization.* New York: Meridian, 1958.

Grant, Campbell, *Rock Art of American Indians.*

Greenberg, Joseph, "Africa as a Linguistic Area." In *Continuity and Change in African Cultures,* edited by William Bascom and Melville Herskovits. Chicago: University of Chicago Press, 1959.

Griaule, Marcel, *Conversations with Ogotemmeli.*

Heider, Karl G., *The Dugum Dani.*

Herodotus of Halicarnassus, *The Histories.*

Hertz, Robert, *Death and the Right Hand.* Translated by Rodney and Claudia Needham. Glencoe, Ill.: The Free Press, 1960.

Karsten, Rafael, *The Headhunters of the Western Amazonas.*

Kenyon, Kathleen, *Archaeology in the Holy Land.* New York: Praeger, 1960.

Kimura, Doreen, "Asymmetry of the Brain." *Scientific American,* March 1973.

Marshak, Alexander, *The Roots of Civilization.*

Meggitt, M. J., *Desert People.*

Newman, Philip L., *Knowing the Gururumba.*

Partridge, Eric, *Origins.* New York: Macmillan, 1961.

Reina, Reuben E., "Cultural Diversity and Behavioral Integration: The Human Skull Ritual Among the Lowland Maya of Northern Guatemala." In *Context and Meaning in Cultural Anthropology*, edited by M. Spiro. New York: The Free Press, 1965.

Seeger, Anthony, "The Meaning of Body Ornaments: A Suya Example."

Snow, Dean R., "Rock Art and the Power of Shamans." *Natural History*, February 1977.

Turner, Victor W., *The Forest of Symbols.*

Waters, Frank, *The Book of the Hopi.* New York: Viking, 1963.

Williams, Thomas Rhys, *The Dusun, a North Borneo Society.* New York: Holt, Rinehart and Winston, 1965.

———, *Introduction to Socialization.* St. Louis: Mosby, 1972.

Zegwaard, Gerald A., "Headhunting Practices of the Asmat of Western New Guinea." *American Anthropologist*, October 1959.

7. DEAD ENDS: *The Myth and Magic of Hair*

Adamson, Joy, *The Peoples of Kenya.*

Adriani, N., and Albert C. Kruyt, *The Bare'e-Speaking Toradja.*

Batchelor, John, *Ainu Life and Lore.*

Becher, Hans, *The Surara and Pakidai: Two Yanoama Tribes in Northwest Brazil.* Vol. 26. Hamburg: Mitteilungen aus dem Museum für Volkerkunde, 1960.

Best, Elsdon, *The Maori.*

Bruemmer, Fred, "The Northernmost People." *Natural History*, February 1974.

Campbell, Bernard G., *Human Evolution.*

Coze, Paul, "Indian Hair-dos." *Arizona Highway*, July 1950.

Densmore, Frances, *Chippewa Customs.*

Díaz del Castillo, Bernal, *The Discovery and Conquest of Mexico.* Translated by A. P. Maudslay. New York: Farrar, Straus and Cudahy, 1956.

Driver, Harold E., *Indians of North America.*

Dubois, J. A., *Hindu Manners, Customs, and Ceremonies.*

Durant, Will, *Our Oriental Heritage.*

Etkin, William, "Social Behavior and the Evolution of Man's Mental Faculties." In *Culture and the Evolution of Man*, edited by Ashley Montague. London: Oxford University Press, 1962.

Evans-Pritchard, E. E., *The Nuer.* Oxford: Clarendon Press, 1940.

Freuchen, Peter, *Book of the Eskimo.* New York: World, 1957.

Goldman, Irving, *The Cubeo.*

Gould, Richard A., *Yiwarra.*

Granet, Marcel, *Chinese Civilization.*

Gulliver, P. H., "The Jie of Uganda." In *Peoples of Africa,* edited by James L. Gibbs, Jr. New York: Holt, Rinehart and Winston, 1965.

Heider, Karl G., *The Dugum Dani.*

Holmberg, Allan R., *Indians of the Long Bow: The Siriono of Eastern Bolivia.* Institute of Social Anthropology publication no. 10, Smithsonian Institution, 1950.

Karsten, Rafael, *The Headhunters of the Western Amazonas.*

Lattimore, Owen, *Manchuria, Cradle of Conflict.* New York: Macmillan, 1935.

Leach, Edmund R., "Magical Hair." In *Myth and Cosmos,* edited by John Middleton. Garden City, N.Y.: Natural History Press, 1967.

Lee, Dorothy, "Greece." In *Cultural Patterns and Technical Change,* edited by Margaret Mead. New York: New American Library, 1955.

Malinowsky, Bronislaw, *The Sexual Life of Savages.* New York: Harcourt, Brace and World, 1929.

McGee, W. J., *The Seri Indians.* Bureau of American Ethnology Annual Report no. 17, part 1, 1895–1896.

Meggitt, M. J., *Desert People.*

O'Malley, L.S.S., *Popular Hinduism.*

———, ed., *Modern India and the West.*

Oswalt, Wendell H., *This Land Was Theirs.*

Singer, Isidore, ed., *The Jewish Encyclopedia,* 1901.

Steward, Julian H., ed., *Handbook of South American Indians.*

Taleyesva, Don, *Sun Chief: The Autobiography of a Hopi Indian.* Translated and edited by Leo W. Simmons. New Haven, Conn.: Yale University Press, 1942.

Titiev, Mischa, *Old Oraibi.*

Tozzer, Alfred M., *Landa's Relacioñ de las Cosas de Yucatân.*

Turner, Terence S., "Tchikrin: A Central Brazilian Tribe and Its Symbolic Language of Bodily Adornment."

Ucko, Peter J., "Penis Sheaths: A Comparative Study." *Proceedings of the Royal Anthropological Institute of Great Britain and Ireland,* 1969.

Washington State University Far Eastern and Russian Institute, *Regional Handbook on Northeast China.* New Haven, Conn.: Human Relations Area File.

8. OF TIME AND BODY: *The Cosmic Connection*

Anisimov, A. F., "Cosmological Concepts of the Peoples of the North." In *Studies in Siberian Shamanism,* edited by H. N. Michaels. Toronto: University of Toronto Press, 1963.

Aveni, Anthony F., ed., *Archaeoastronomy in Precolumbian America.* Austin: University of Texas Press, 1975.

———, ed., *Native American Astronomy.* Austin: University of Texas Press, 1977.

Balog, Janos, "Blackouts and Births." *Natural History,* April 1978.

Berndt, Ronald M., *Excess and Restraint.*

——— and Catherine H. Berndt, *The World of the First Australians.*

——— and ———, eds., *Aboriginal Man in Australia.* Sydney: Angus and Robertson, 1965.

Bogoras-Tan, Vladimir Germanovich, *The Chukchee.*

Coe, Michael D., "Native Astronomy in Mesoamerica." In *Archaeoastronomy in Precolumbian America,* edited by A. Aveni.

Driver, Harold E., *Indians of North America.*

Eddy, John A., "Archaeoastronomy of North America." In *In Search of Ancient Astronomies,* edited by E. C. Krupp. New York: Doubleday, 1977.

———, "Medicine Wheels and Plains Indian Astronomy." In *Native American Astronomy,* edited by A. Aveni.

Elkin, A. P., *The Australian Aborigines.*

Ellis, Florence Hawley, "A Thousand Years of the Pueblo Sun-Moon-Star Calendar." In *Archaeoastronomy in Precolumbian America,* edited by A. Aveni.

Erikson, Erik H., *Observations on the Yurok.* University of California Publications in Archaeology and Ethnology, vol. 35, no. 10, 1943.

Evans, Joan, *Time and Chance.* London: Longmans, 1943.

Frisch, Rose E., and Janet W. McArthur, "Menstrual Cycles: Fatness as a Determinant of Minimum Weight for Height Necessary for Their Maintenance or Onset." *Science,* September 13, 1974.

Goldman, Irving, *The Cubeo.*

Graves, Robert, *The Greek Myths.* New York: Braziller, 1957.

Gusinde, Martin, *The Yahgan.*

Heider, Karl G., *The Dugum Dani.*

Herodotus of Halicarnassus, *The Histories.*

Karsten, Rafael, *The Headhunters of the Western Amazonas.*

———, *The Religion of the Samek.* Leiden: E. J. Brell, 1955.

Kolata, Gina Bari, "!Kung Hunter-Gatherers: Feminism, Diet, and Birth Control." *Science,* September 13, 1974.

Kroeber, Alfred E., "Yurok Law and Custom." In *The California Indians,* edited by Heizer and Whipple.

Kroeber, Theodora, *The Inland Whale.* Berkeley: University of California Press, 1971.

Krupp, E. C., ed., *In Search of Ancient Astronomies.* New York: Doubleday, 1977.

Leroi-Gourhan, André, *Treasures of Prehistoric Art.*

Malinowsky, Bronislaw, *The Sexual Life of Savages.*

Marinatos, Spyridon, *Crete and Mycenae.* New York: Abrams, 1960.

Marshak, Alexander, "Upper Paleolithic Notation and Symbol."

———, *The Roots of Civilization.*

———, "An Olmec Mosaic Pendant." In *Archaeoastronomy in Precolumbian America,* edited by A. Aveni.

Mead, Margaret, *Growing Up in New Guinea.* New York: New American Library, 1953.

Mellaart, James, *Çatal Hüyük.*

———, *The Neolithic of the Near East.*

Orlovskii, P. N., "A Year in the Life of an Anadyr-Chukchee Reindeer Raiser." *Severnaia Aziia,* Kn. 2 (1928): 61–70.

Palau Marti, Monserrat, *The Dogon.*

Paulme, Denise, *Social Organization of the Dogon.*

Radcliffe-Brown, A. R., *The Andaman Islanders.*

Remington, Judith, "Current Astronomical Practices Among the Maya." In *Native American Astronomy,* edited by A. Aveni.

Schebesta, Paul, *The Negritos of Asia.* Translated by Frieda Schultze. New Haven, Conn.: HRAF Press, 1962.

Shternberg, Lev, *The Gilyak, Orochi, Goldi, Negidol, Ainu: Articles and Materials.* Khaharovsk, Dal'giz, 1933. 39. Edited and preface by I. A. P. Allkor (Koshkin).

Strathern, Andrew, and Marilyn Strathern, *Self Decoration in Mt. Hagen.*

Sverdrup, Harald Ulrik, *With the People of the Tundra.* Oslo: Gyldendal Norsk Forlag, 1938.

Tozzer, Alfred M., *Landa's Relación de las Cosas de Yucatân.*

Turnbull, Colin, *The Forest People.* New York: Simon and Schuster, 1961.

Van Beek, Gus W., "The Rise and Fall of Arabia Felix." *Scientific American,* December 1969.

Veniamenov, Ivan Evsieevich Popov, *Notes on the Islands of the Unalaska District.* St. Petersburg: Russian-American Co., 1840.

9. DENYING THE FLESH: *In Command of the Appetites*

Arens, William, "The Great Football Ritual."

Blunt, Edward, ed., *Social Service in India.* London: His Majesty's Stationery Office, 1938.

Diener, Paul, and Eugene E. Robkin, "Ecology, Evolution, and the Search for Cultural Origins: The Question of Islamic Pig Prohibition." *Current Anthropology,* March 1978.

Downs, James F., *The Two Worlds of the Washo.* New York: Holt, Rinehart and Winston, 1966.

Driver, Harold E., *The North American Indians.*

Durant, Will, *Our Oriental Heritage.*

———, *The Age of Faith.* New York: Simon and Schuster, 1950.

Firth, Raymond, *Tikopia Ritual and Belief.* Boston: Beacon Press, 1967.

Goldman, Irving, *The Cubeo.*

———, *Ancient Polynesian Society.* Chicago: University of Chicago Press, 1970.

Goldschmidt, Walter, "The Ethical Prescriptions of Yurok Society." In *Exploring the Ways of Mankind,* edited by W. Goldschmidt. New York: Holt, Rinehart and Winston, 1977.

Hallowell, A. Irving, *Culture and Experience.* Philadelphia: University of Pennsylvania Press, 1955.

Harlan, Edgar R., et al., *An Original Study of Mesquakie Life.* Annals of Iowa, Third Series, 19. 1933/1935.

Harrington, J. H., ed., *Catholic Encyclopedia.* New York: McGraw-Hill, 1965.

Harris, Marvin, "The Myth of the Sacred Cow." *Natural History,* March 1967.

———, "The Cultural Ecology of India's Sacred Cattle." *Current Anthropology,* February 1966.

———, "The Riddle of the Pig I." *Natural History,* October 1972.

———, "The Riddle of the Pig II." *Natural History,* February 1973.

Herodotus of Halicarnassus, *The Histories.*

Hoebel, E. Adamson, *The Cheyennes.* 2d ed. New York: Holt, Rinehart and Winston, 1978.

Jenness, Diamond, *The Ojibwa of Parry Island.* Ottawa: Department of Mines, National Museum of Canada Bulletin no. 78, 1935.

Jones, William, *Ethnography of the Fox Indians,* Edited by Margaret Welpley Fisher. Bureau of American Ethnology Bulletin no. 125, 1939.

Karve, Irawati, *Kinship Organization in India.* Poona: Deccan College Monograph Series, 1953.

Kraft, Ken, and Pam Kraft, "Secrets of the Great Plant Wizard." *Natural History,* October 1973.

Kroeber, Alfred E., "Yurok." In *The Languages of the Coast of California North of San Francisco.* Berkeley: University of California Press, 1911.

Landes, Ruth, *The Ojibwa Woman.* New York: Columbia University Press, 1938.

Michelson, Truman, "The Traditional Origin of the Fox Society Known as 'The Singing Around Rite.' " Bureau of American Ethnology Annual Report no. 40, 1918–1919.

Neumann, Thomas W., "A Biocultural Approach to Salt Taboos: The Case of the Southeast United States." *Current Anthropology,* June 1977.

O'Malley, L.S.S., *Popular Hinduism.*

Ortner, Sherry B., "Sherpa Purity." *American Anthropologist,* February 1963.

Paulme, Denise, *Social Organization of the Dogon.*

Rasmussen, Knud, *Report of the Fifth Thule Expedition, 1921–24.* Vol. 7, no. 1. Copenhagen: Gyldendalske Boghandel, Nordisk Forlag, 1929.

Ross, Eric Barry, "Food Taboos, Diet, and Hunting Strategy: The Adaptation to Animals in Amazon Cultural Ecology." *Current Anthropology,* March 1978.

Sahagún, Bernardino de, *Florentine Codex.*

Scotch, Norman A., "Magic, Sorcery, and Football Among Urban Zulu: A Case of Reinterpretation Under Acculturation." In *Black Africa,* edited by John Middleton. New York: Macmillan, 1970.

Singer, Isidore, ed., *The Jewish Encyclopedia.*

Stevenson, Margaret, *The Rites of the Twice Born.*

Swanton, John Reed, *Social Organization and Social Usages of the Indians of the Creek Confederacy.* Bureau of American Ethnology Annual Report no. 40, 1924–1925.

Thomas, Elizabeth Marshall, *The Harmless People.* New York: Knopf, 1959.

Thompson, J. Eric, *Ethnology of the Mayas of Southern and Central British Honduras.* Chicago: Field Museum of Natural History, 1930.

Turner, Victor W., *The Forest of Symbols.*

Villa Rojas, Alfonso, *The Maya of East Central Quintana Roo.* Translated by Burton Lifschutze et al. Washington, D.C.: Carnegie Institution of Washington, 1945.

Waugh, Frederick Wilkerson, Jr., *Foods and Food Preparation of the Iroquois.* Ottawa: Government Printing Bureau, 1916.

Webb, Malcolm C., "Abolition of Taboo System in Hawaii." In *Polynesia,* edited by Alan Howard. Scranton, Pa.: Chandler, 1971.

Weber, Max, *The Protestant Ethic and the Spirit of Capitalism.* London: George Allen and Unwin, 1930.

10. MORTIFYING THE FLESH: *The Sanctity of Pain*

Baring-Gould, S., *The Lives of the Saints.* Edinburgh: John Grant, 1914.

Brain, James L., "Sex, Incest and Death: Initiation Rites Reconsidered." *Current Anthropology,* June 1977.

Carneiro, Robert L., "Hunting and Hunting Magic Among the Amahuaca of the Peruvian Montaña." *Ethnology,* October 1970.

Catlin, George, *O-kee-pa.*

Crooke, W., *The Popular Religion and Folklore of Northern India.* Vol. 1. Westminster, England: Archibald Constable, 1896.

Dorsay, George Amos, *The Arapaho Sun Dance.* Millwood, N.Y.: Kraus Reprint, 1973.

——— and A. L. Kroeber, *Traditions of the Arapaho.* Chicago: Field Columbian Museum, 1903.

Durant, Will, *The Age of Faith.*

Elkin, A. P., *The Australian Aborigines.*

Freeman, James M., "Trial by Fire." *Natural History,* January 1974.

Goldman, Irving, *The Cubeo.*

———, *Ancient Polynesian Society.*

Goodenough, Ward, "The Pageant of Death in Nakanai." University of Pennsylvania Museum Bulletin no. 19, 1955.

Gould, Richard A., *Yiwarra.*

Heider, Karl G., *The Dugum Dani.*

Herodotus of Halicarnassus, *The Histories.*

Hoebel, E. Adamson, *The Cheyennes.*

Jenkinson, Michael, "The Glory of the Long Distance Runner." *Natural History,* January 1972.

Kosambi, D. D., "Living Prehistory in India." *Scientific American,* February 1967.

Meggitt, M. J., *Desert People.*

Nabokov, Peter, *Two Leggings: The Making of a Crow Warrior.* New York: Crowell, 1967.

O'Malley, L.S.S., *Popular Hinduism.*

———, *Modern India and the West.*

Pettitt, George A., "The Vision Quest and the Guardian Spirit." In *Primitive Education in North America.* University of California Publications in American Archaeology and Ethnology, vol. 43, no. 1, 1946.

Schebesta, Paul, *The Negritos of Asia.*

Snyder, Solomon H., "Opiate Receptors and Internal Opiates." *Scientific American,* March 1977.

Tozzer, Alfred M., *Landa's Relaciõn de las Cosas de Yucatân.*

Warner, Esther, *Trial by Sasswood.* London: Gollancz, 1955.

Williams, Thomas Rhys, *The Dusun.*

———, *Introduction to Socialization.*

Wilson, Monica, *Good Company.* Boston: Beacon Press, 1963.

11. USES OF THE FLESH: *Fellow Feeling, Fellow Feeding*

Adriani, N., and Albert C. Kruyt, *The Bare'e-Speaking Toradja.*

Alkema, B., and T. J. Bezemer, *Concise Handbook of the Ethnology of the Netherlands East Indies.* Haarlem: H. D. Tjeenk Willink en Zoon, 1927.

Artamonov, M. I., "Frozen Tombs of the Scythians." *Scientific American,* May 1965.

Berndt, Ronald M., *Excess and Restraint.*

Best, Elsdon, *The Maori.*

Buck, Peter H., *The Coming of the Maori.*

Chang, K. C., *The Archaeology of Ancient China.* New Haven, Conn.: Yale University Press, 1968.

Coe, Michael D., *The Maya Scribe and His World.*

———, "The Shadow of the Olmecs."

———, Personal communication.

Díaz del Castillo, Bernal, *The Discovery and Conquest of Mexico.*

Downs, Richard Erskine, *The Religion of the Bare'e-Speaking Toradja.*

Downs, T. W., "Maori Etiquette." *Polynesian Society Journal* 38, 1929.

Durant, Will, *Our Oriental Heritage.*

Elkin, A. P., *The Australian Aborigines.*

Emery, Walter B., "The Tombs of the First Pharaohs." *Scientific American,* July 1957.

Evans-Pritchard, E. E., *Witchcraft, Oracles and Magic Among the Azande.* Oxford: Clarendon Press, 1937.

Glasse, Robert, "Cannibalism in the Kuru Region of New Guinea." *Transactions of the New York Academy of Science,* series 2, no. 29, 1967.

Glob, P. V., *The Bog People.* Translated by Robert Brace-Mitford. Ithaca: Cornell University Press, 1969.

Goldman, Irving, *Ancient Polynesian Society.*

Graffen, Enrico, and Edoardo Columbo, *The Niam-Niam. Revue Internationale de Sociologie,* 114, 1906.

Granet, Marcel, *Chinese Civilization.*

Harner, Michael J., "The Enigma of Aztec Sacrifice." *Natural History,* April 1977.

Herodotus of Halicarnassus, *The Histories.*

Hertz, Robert, *Death and the Right Hand.*

Kenna, Margaret E., "Houses, Fields, and Graves; Property and Ritual Obligations on a Greek Island." *Ethnology,* January 1976.

Koch, Klaus-Friedrich, "Cannibalistic Revenge in Jale Warfare." *Natural History,* February 1970.

Kozak, Vladimir, "Ritual of a Bororo Funeral." *Natural History,* January 1963.

Lawson, *Modern Greek Folklore and Ancient Greek Religion.* Cambridge, England: Cambridge University Press, 1910.

Le Page de Pratz, Antoine S., *The History of Louisiana.* 3 vols. Paris, 1758.

Léry, Jean de, *History of a Voyage to Brazil.* Edited by Paul Gaffarel. Paris: Alfonse Lemerre, 1880.

Mellaart, James, *Çatal Hüyük.*

Metcalf, Peter A., "Death Be Not Strange." *Natural History,* June/July 1978.

Moorey, P.R.S., "What Do We Know About the People Buried in the Royal Cemetery?" *Expedition,* Fall 1977.

Mostny, Grete, "The Boy from Lead Mountain." In *History Was Buried,* edited by Margaret Wheeler. New York: Hart Publishing Co., 1967.

Muller, Kai, "Taboos and Magic Rule Namba Lives." *National Geographic,* January 1972.

Ortiz de Montellano, Bernard R., "Aztec Cannibalism: An Ecological Necessity?" *Science,* May 12, 1978.

Phillips, E. D., "The Nomad People of the Steppes." In *The Dawn of Civilization,* edited by Stuart Piggott. New York: McGraw-Hill, 1961.

Sahagún, Bernardino de, *Florentine Codex.*

Singer, Isidore, ed., *The Jewish Encyclopedia.*

Staden, Hans, *The True Story of His Captivity.* Translated and edited by Malcolm Letts. London: Routledge and Sons, 1928.

Ssu-ma Ch'ien, *Records of the Grand Historian of China.* 2 vols. Translated from the *Shih Chi* and arranged by Burton Watson. New York: Columbia University Press, 1961.

Topping, Audrey, "The First Emperor's Army: China's Incredible Find." *National Geographic,* April 1978.

Tozzer, Alfred M., *Landa's Relación de las Cosas de Yucatán.*

Trigger, Bruce G., *The Hurons.* New York: Holt, Rinehart and Winston, 1969.

Willett, Frank, *Ife in the History of West African Sculpture.* New York: McGraw-Hill, 1967.

Woolley, Leonard, *Excavations at Ur.* London: Ernest Benn, 1954.

12. THROUGH FLESH TO FANTASY: *Sacred Senses*

Aberle, David F., *The Peyote Religion Among the Navaho.* Viking Fund Publications in Anthropology no. 42, 1966.

Artamonov, M. I., "Frozen Tombs of the Scythians."

Bogoras-Tan, Vladimir Germanovitch, *The Chukchee.*

Carneiro, Robert L., "Hunting and Hunting Magic Among Amahuaca of the Peruvian Montaña."

Castaneda, Carlos, *The Teachings of Don Juan.* Berkeley: University of California Press, 1968.

Chagnon, Napoleon A., *Yąnomamö: The Fierce People.*

———, Philip Le Quesne, and James M. Cook, "Yąnomamö Hallucinogens: Anthropological, Botanical, and Chemical Findings." *Current Anthropology,* February 1977.

Coe, Michael D., "The Shadow of the Olmecs."

———, Personal communication.

Cooper, John M., "Stimulants and Narcotics." In *Handbook of South American Indians,* vol. 5.

Dobkin de Rios, Marlene, "The Influence of Psychotropic Flora and Fauna on Maya Religion." *Current Anthropology,* June 1974.

Dunn, Ethel, "Russian Use of *Amanita Muscaria*: A Footnote to Wasson's Soma." *Current Anthropology,* October 1973.

Driver, Harold E., *The North American Indians.*

Embolden, William A., Jr., "Ritual Uses of *Cannabis Sativa L.*" In *Flesh of the Gods,* edited by Peter T. Furst.

Fuchs, Andrew, "Coca Chewing and High Altitude Stress: Possible Effects of Coca Alkaloids on Erythropoiesis." *Current Anthropology,* June 1978.

Furst, Peter T., ed., *Flesh of the Gods.* New York: Praeger, 1972.

———, "To Find Our Life: Peyote Among the Huichol Indians of Mexico." In *Flesh of the Gods.*

———, "An Indian Journey to Life's Source." *Natural History,* April 1973.

——— and Michael D. Coe, "Ritual Enemas." *Natural History,* March 1977.

Goldman, Irving. *The Cubeo.*

———, Personal communication.

Grant, Campbell, *Rock Art of American Indians.*

Hall, Robert L., "An Anthropocentric Perspective for Eastern United States Prehistory." *American Antiquity,* October 1977.

Hanna, Joel M., "Coca Leaf Use in Southern Peru: Some Biosocial Aspects." *American Anthropologist,* June 1974.

Harner, Michael J., ed., *Hallucinogens and Shamanism.* London: Oxford University Press, 1973.

———, "The Sound of Rushing Water." In *Hallucinogens and Shamanism.*

———, "The Role of Hallucinogenic Plants in European Witchcraft." *In Hallucinogens and Shamanism.*

———, "Common Themes in South American Indian *Yage* Experiences." In *Hallucinogens and Shamanism.*

Heizer, R. F., and M. A. Whipple, *The California Indians.*

Herodotus of Halicarnassus, *The Histories.*

Kaplan, Reid W., "The Sacred Mushroom in Scandinavia." *Man,* March 1975.

Lee, Richard B., "Trance Cure of the !Kung Bushman." *Natural History,* November 1967.

Magical Death, A film about the Yąnomamö.

McAllister, J. Gilbert, *Daveko: Kiowa-Apache Medicine Man.* Bulletin 17 of the Texas Memorial Museum. Austin: University of Texas Press, 1970.

Munn, Henry, "The Mushrooms of Language." In *Hallucinogens and Shamanism,* edited by Harner.

Naranjo, Claudio, "Psychological Aspects of the *Yage* Experience in an Experimental Setting." In *Hallucinogens and Shamanism,* edited by Harner.

Sahagún, Bernardino de, *Florentine Codex.*

Schultes, Richard E., "An Overview of Hallucinogens in the Western Hemisphere." In *Flesh of the Gods,* edited by Furst.

———, "Man and Marijuana." *Natural History,* August–September 1973.

Siegel, Ronald K., "Hallucinations." *Scientific American.* October 1977.

Siskind, Janet, "Visions and Cures Among the Sharanahua." In *Hallucinogens and Shamanism*, edited by Harner.

Snow, Dean, "Rock Art and the Power of Shamans."

Tozzer, Alfred M., *Landa's Relación de las Cosas de Yucatán.*

Wellmann, Klaus F., "North American Rock Art and Hallucinogenic Drugs." *Journal of the American Medical Association*, April 14, 1978.

Wenner, Kate, "The Farm." *New York Times Magazine*, May 8, 1977.

EPILOGUE: *Body and Soul*

Forde, Daryll, ed., *African Worlds.* London: Oxford University Press, 1954.

Goldman, Irving, *The Mouth of Heaven.*

Griaule, Marcel, and Germaine Dieterlen, "The Dogon." In *African Worlds*, edited by Daryll Forde.

Jenness, Diamond, *The Ojibwa of Parry Island.*

Marsh, Gordon H., *A Comparative Survey of Eskimo-Aleut Religion.* Anthropological Papers of the University of Alaska 3, 1954.

Mercier, P., "The Fon of Dahomey." In *African Worlds*, edited by Daryll Forde.

Oswalt, Wendell, *Alaskan Eskimos.* San Francisco: Chandler, 1967.

Stanner, W.E.H., "The Dreaming." In *Australian Signpost*, edited by T.A.G. Hungerford. Melbourne: F. W. Cheshire Publishing Pty., Ltd., 1956.

INDEX
OF
PEOPLES AND PLACES

H

I

J

K

L

M

N

O

P